DEEP SEISMIC SOUNDING OF THE EARTH'S CRUST AND UPPER MANTLE

METOD GLUBINNOGO SEISMICHESKOGO ZONDIROVANIYA ZEMNOI KORY I VERKHOV MANTII

МЕТОД ГЛУБИННОГО СЕЙСМИЧЕСКОГО ЗОНДИРОВАНИЯ ЗЕМНОЙ КОРЫ И ВЕРХОВ МАНТИИ

DEEP SEISMIC SOUNDING OF THE EARTH'S CRUST AND UPPER MANTLE

Irina P. Kosminskaya

O. Yu. Shmidt Institute of Terrestrial Physics
Moscow, USSR

Translated from Russian by

George V. Keller
Colorado School of Mines
Golden, Colorado

SPRINGER SCIENCE+BUSINESS MEDIA, LLC · 1971

The original Russian text, published for the O. Yu. Shmidt Institute of
Terrestrial Physics by Nauka Press in Moscow in 1968, has been corrected
by the author for the present edition. The English translation is published
under an agreement with Mezhdunarodnaya Kniga, the Soviet book export
agency.

Косминская И. П.

Метод глубинного сейсмического зондирования земной коры и верхов мантии

Library of Congress Catalog Card Number 79-131887

ISBN 978-1-4684-7869-3   ISBN 978-1-4684-7867-9 (eBook)
DOI 10.1007/978-1-4684-7867-9

© 1971 Springer Science+Business Media New York
Originally published by Consultants Bureau, New York in 1971

INTRODUCTION

Deep seismic sounding was proposed by G. A. Gamburtsev and developed under his guidance during the period 1948-1955 at the Institute of Physics of the Earth of the Academy of Sciences of the USSR. During that period also, the first geophysical results concerning the deep structure of the earth's crust in several regions in Tien-Shan, the Pamir, and Turkmenia were observed.

Beginning with 1956, the deep seismic sounding method has been used widely by geophysical research groups as well as by geophysical service organizations for regional studies in the USSR. Descriptions of this work have been given in reports by Yu. N. Godin, V. V. Fedynskii, D. N. Kazanli, and others. New variants of the deep seismic sounding method have been developed; continuous profiling (Yu. N. Godin, and others), and point soundings (N. N. Puzyrev, and others). Deep seismic soundings have been carried on outside of Russia also, and studies have been carried out on the use of the deep seismic sounding method in marine applications (E. I. Gal'perin, S. M. Zverev, I. P. Kosminskaya, Yu. P. Neprochnov, and others).

Over the past decade, the deep seismic sounding method has joined the suite of geophysical studies as a highly detailed method for studying the earth's crust and upper mantle to depths of 50 to 100 km on land, and of 15 to 25 km in the deep oceans.

Using the deep seismic sounding method, deep structure has been studied in a variety of tectonic provinces of the USSR, in the basins of the Black, Caspian, Okhotsk, and Japan seas, and in the Kuriles – Kamchatka portion of the Pacific Ocean; studies have also been carried out in the Indian Ocean. In recent years, the deep seismic sounding method has come to be used quite widely outside the USSR as well; in the USA, in the form of segmentally continuous profiling, and in western Europe in both the continuous and segmentally continuous profiling forms.

Deep seismic sounding investigations as part of the Internation Upper Mantle Project are serving as the basis for several international geophysical efforts, which include a project for studying the deep structure of the Carpathians-Balkans region in several neighboring Socialist countries.

Development of the deep seismic sounding method has been based on exploration seismology technology. In turn, deep seismic sounding developments have had their effect on methods used in seismic exploration by increasing the depths which are studied, leading to the use of regional surveys in mining exploration, resulting in the development of several aspects of the theory of wave propagation, and so on. The combination of standard seismic exploration with deep seismic sounding has allowed detailed studies of the upper part of the crystalline crust and selection of sites suitable for extra deep drilling (the Baltic Sheet, Kurin Lowlands, and others).

During gathering of data on the crystal structure in various tectonic regions, some of the fundamentals of deep seismic sounding have been developed. The main features of these studies are given in papers by Yu. N. Godin and his colleagues – I. S. and B. S. Vol'vovskii, I. V. Pomerantseva, A. V. Egorkin, V. Z. Ryaboi, and others; by D. N. Kazanli and his colleagues – A. A. Popov, A. P. Antonenko, and others; by I. V. Litvinenko et al.; V. B. Sollogub, V. A. Chekunova, and others; Yu. P. Neprochnov and others; N. N. Puzyrev; and S. B. Krylov and others; as well as in papers describing technique from the Institute of Physics of the Earth by E. I. Gal'perin, S. M. Zverev, P. S. Veitsman, Yu. V. Tulina, N. I. Davydov, G. G. Mikhota, et al. Papers by the author on methods of analyzing wave fields and the geological evaluation of deep seismic sounding data represent a recent stage in the work. Fundamental studies of the propagation of waves under deep seismic sounding conditions have been carried out by A. S. Alekseev.

The present book is a review of studies in the area of development of the basis for the method. I have examined the question of the physical basis for optimum frequency bands for recording on land or at sea, of the resolution of the deep seismic sounding method for various field procedures, of the systematic analysis of wave fields, and of the geological interpretation of deep seismic sounding data. The book is based on experimental data obtained in part by the author in marine and land-based deep seismic sounding operations of the Institute of Physics of the Earth of the Academy of Sciences, USSR, as well as from numerous Russian and non-Russian publications and the personal awareness of the author of primary deep seismic sounding data from many regions of the USSR.

Some of the material used in this book was obtained by the author's deep seismic group of the Physics of the Earth Institute, and has not been published previously. This includes spectral studies carried out by G. G. Mikhota, deep seismic sounding modelling done by A. G. Aver'yanov, work by S. M. Zverev on development of methods for studying the upper mantle under the oceans, investigations by Yu. V. Tulina on the analysis of structural anisotropy in the Southern Kuriles, work by A. N. Fursov, N. I. Davydov, and G. A. Yaroshevskii on the study of near-critical reflections with magnetic recording equipment, and a number of others.

All of the calculations which are given in the tables and graphs were done by E. N. Zaitseva's engineering group. The illustrations were done by Z. D. Babaeva, I. S. Lysova and I. V. Sokratova. Many of the sections were constructed by G. A. Krasil'shchikova.

I express my thanks to all of my friends who assisted in selection of the material and preparation of the manuscript, which was prepared for publication by S. I. Masarskii.

The author acknowledges the assistance of A. S. Alekseev, V. I. Keilis-Borok, T. B. Yanovskaya, N. S. Smirnova, V. T. Cherveni, and G. S. Pod"yapol'sky on theoretical aspects.

The author extends thanks to V. V. Belousov, Yu. M. Krestnikov, A. A. Borisov, I. A. Rezanov, and G. B. Udintsev for ideas which they gave freely in explaining many aspects of the geological meaning of deep seismic sounding data.

The author wishes to extend her thanks also to colleagues in the Deep Seismic Sounding Section – I. S. Vol'vovskii, I. V. Litvinenko, Yu. B. Demidenko, V. B. Sollogub, N. I. Davydov, N. N. Puzyrev, P. S. Veitsman, V. Z. Ryaboi, A. A. Popov, A. V. Egorkin, and N. I. Khalevin, with whom I had many fruitful discussions.

CONTENTS

CHAPTER I

DEEP SEISMIC SOUNDING AND ITS RELATION TO OTHER SEISMIC METHODS

Over the last decade, a trend has become apparent in the development of geophysics — a trend toward planetary-scale studies, averaged to a large degree, in the search for a basic cause for regional and local geological behavior. This trend is represented not only by the use of already abundant data on the general structure of the earth, but also by the attainment of new levels in the development of methods and techniques for geophysical investigations, which now allow a more detailed study of the Earth's structure. Deep seismic sounding is one of these new detail methods. Its appearance among the older seismic methods has made possible more specific studies of the crust.

A second feature of the present-day overall development of geophysics seems to be the integration of geophysics with the other earth sciences, primarily with the other branches of geology. This last trend is especially apparent in the area of studies of the earth's crust.

§ 1. Development of Crustal Studies and Deep Seismic Sounding

The following three periods may be recognized in the history of the study of the earth's crust and upper mantle.

The first period dates from the discovery in 1909 by Mohorovičic of a seismic boundary which has come to be accepted as the lower boundary of the earth's crust. Over the ensuing thirty years, seismologists all over the world studied the existence of this boundary, using for this purpose seismograms from a seismograph network around the world which was sparse and unevenly distributed. It is characteristic of this period that only indifferent attempts were made to relate the geophysical studies to geology, and the physical nature of the earth's interior was considered as some abstract concept, quite divorced from geological ideas.

The second period of development, during the forties and fifties, is represented by the time when a number of seismologists (Gutenberg, Caloi, Rizova, and others) attempted to represent the crust in a number of regions on the basis of earthquake records obtained from a more limited network of seismographs in a single region. It was during this interval that Gutenberg advanced his hypothesis about mountain building. His ideas were of great interest to geologists. In these, he first pointed out the relationship between the deep structure of the earth and features which are quite obvious at the surface. However, even during this second period, the relationship between geology and geophysics was still quite nebulous and not very satisfactory.

During this same period, in other areas of seismology — seismic prospecting — the integration between geology and geophysics was taking place because of the need in specific ex-

1

ploration problems, particularly in the search for oil and gas. Geological exploration incorporated as one of its tools, geophysics, particularly seismic prospecting, as an integral part of the exploration effort.

The beginning of the third and contemporaneous period in the development of crustal studies is associated with the appearance and wide use of the deep seismic sounding method which has allowed not only the determination of averaged crustal structure over large, complicated regions, but also the construction of seismic sections along specifically located seismic profiles in regions with various geological structures and various histories of geological development. The results from this new stage in crustal studies have played a major role in the incorporation of geological concepts into geophysics [14–19, and others]. Geology has not only obtained data on the deep structure of the earth in the usual form of sections but also the possibility of comparing these with specific well-known geological features, and not with such diluted concepts as the "average crust of Europe" or the "crust of central Asia."

At this time, the rewarding cooperation between geologists and geophysicists in the study of the earth's crust started to take place. Soviet researchers have played a leading role in developments along these lines, starting with the work of Yu. N. Godin, who first used deep seismic sounding data in Turkmenia for studying broad tectonic features. He was also the first (in 1952) to point out the importance of information on deep crustal structure in exploration programs for predicting the occurrence of oil and gas fields, ore deposits and other features, so that the deep seismic sounding method was incorporated in geological exploration programs throughout the Soviet Union.

The depth of the relationship between geology and geophysics is particularly apparent in the explanation of the specific crustal structures in various tectonic zones and the search for an explanation of their evolution. The introduction of geology into the interpretation of deep geophysical data leads to the recognition of new categories in deep studies. For our purposes, we need only note here that in the interpretation of deep seismic soundings, not only are geological data such as sections, maps, and structural representations used in interpretation, but also the combination of all geophysical and geological studies which can contribute to an understanding of the geological history of the region under study.

§ 2. The Role of Deep Seismic Sounding among

Other Seismic Methods

In addition to the development of investigations of the earth's crust and upper mantle and with the acquisition of information about its structure, various forms of seismic studies have been developed for use in these studies. Let us now consider the role of deep seismic sounding among these seismic methods.

Present-day seismology is the result of studies related to vibrations of the solid earth under the action of impulsive excitation: earthquakes and explosions. The study of persistent, quasistationary oscillations (microseisms) plays a subordinate role in seismology; long-period quasistationary oscillations of the solid earth — resonant modes — are studied by gravimetry.

A primary purpose of seismology is the study of earth structure, and the best data to use for such a purpose are obtained from recording vibrations generated by an impulsive source. At present, various types of waves from an explosion or earthquake are used, being recorded over a broad range of frequencies (nearly five decades), from less than one hertz to tens per minute. This allows the study of the earth in general and its zonal structure with varying detail.

Seismic investigations directed to the study of earth structure might be subdivided into two major categories, depending upon the type of source that generates the waves. With this approach, we may speak of earthquake seismology and explosion seismology.

Earthquake Seismology. Systematic recording of reasonably strong earthquakes (as well as large explosions) is done with a worldwide network of seismograph observatories, and these data have been used to determine the internal structure of the earth in general, and to define the velocity profile from the surface to the center, through both the mantle and core (Jeffries, Gutenberg, Bullen). However, this network is sparse and not evenly distributed over the Earth's surface; the spacing between stations is smaller in the economically advanced areas and in seismically active areas, and less dense in the underdeveloped areas and aseismic areas (which frequently do not coincide). There are very few seismic stations over extensive oceanic expanses. The data from widely separated observatories cannot be used to provide a picture of the structure in detail in these regions. Evidence of this type is obtained from data from denser regional networks of recording stations or from other, more specialized survey procedures.

Regional seismic data have been used to determine the general character of the structure of the earth's crust and upper mantle in many of the seismically active areas (Gutenberg, Leman, Caloi, Rozova). Improvements in what may be done with regional seismology with respect to more detailed study of earth structure take place from year to year as the density of the seismograph network is systematically increased [156]. The original ideas about the general structure of the earth on a planetary scale — major shells with the use of the principal epicentral zones of the earth — have been confirmed by these studies (Riznochenko [148]).

Regional travel-time curves for body waves have been constructed for many of the seismically active areas, using data accumulated over many years. The amount of data required to determine regional travel-time curves is also adequate to study planetary-scale inhomogeneities in the mantle. The precision of these data may now be substantially improved by introducing corrections for inhomogeneities in crustal structure which may be determined from special surveys.

However, evidence about the zonal structure of the earth, and particularly of the crust and mantle, which may be obtained from this seismograph network cannot provide satisfactory answers to many of the questions which have arisen in modern earth science, including seismology itself.

Therefore, over the past decade, detailed studies of earthquakes such as were begun by G. A. Gamburtsev in the fifties, and called experimental seismology by him, have become of primary importance [50].

This type of investigation is represented by the operations of the combined Seismological Field Team of the Institute of Physics of the Earth [119, 127] in the study of regional seismicity. New methods for studying the structure of the crust and upper mantle using body waves from earthquakes recorded on high-sensitivity equipment have been developed by this field team. The data so obtained may be termed detailed seismology.

Detailed seismic observations from equipment recording simultaneously along a profile are of particular interest, and are termed "seismic profiling."

Explosion Seismology. Explosion seismology has developed only since the twenties of this century and has been applied primarily as an exploration tool.

Now, explosion seismology includes not only local seismic exploration but also the study of the propagation of waves from an explosion to considerable distances from the shotpoint. In particular, this branch of seismology includes efforts directed toward the study of the upper crust (regional seismic exploration to a depth of 5–10 km) and specialized procedures to survey the crust through its entire thickness – deep seismic sounding – which have been used more recently to study the upper mantle as well, to depths of 80–100 km. Deep seismic sounding fits in, on the one hand, with deep regional seismic exploration, and on the other, with detailed and regional seismology, particularly seismic profiling. The study of earthquake surface waves generated by impulsive or time-stationary sources (microseisms) is a less detailed method of study than is the deep seismic sounding method. This approach is based on the study of dispersion of surface waves.

Of the many field techniques used in deep seismic sounding, which differ markedly in detail, techniques which provide observations in enough detail to permit identification and tracing of individual events by correlation have an advantage. Such a technique is the observation of arrivals from large explosions using a large number of detectors arrayed along a profile.

This outline indicates that contemporary seismology can make use of a broad range of techniques for the study of the internal structure of the earth, with information obtained from earthquake records from networks of stationary and temporary recording stations, as well as from special surveys using explosions. This information varies significantly with respect to adequacy and detail because of the limitations of the various methods as well as because of the practical limitations to the density of observations that can be obtained. For example, considering the physical causes for body and surface waves, it is obvious that the body waves will provide greater precision in the study of an inhomogeneous medium. However, this comparison will not hold if the body waves are observed only at a limited number of stations or along a short profile. If the earth happens to be horizontally layered, observations of the dispersion of quasi-stationary surface waves over a short interval in distance will give results which are sometimes comparable in accuracy with those from impulsive body waves, despite the fact that the body waves usually characterize the medium more accurately and in greater detail.

§ 3.   Primary Methods of Studying the Crust
and Upper Mantle

Thus, in the study of the crust and upper mantle, the methods used most widely are the surface-wave method, detailed seismological studies along profiles of recording stations, and the deep seismic sounding method.

We may characterize the specifications for these methods in greater detail, from the point of view of their applicability for the solution of regional problems: the study of the crust and mantle with sufficient detail so as to permit reasonable differentiation of the underlying controls existing in grossly different tectonic zones of the continents and oceans. Deep structure in such zones is distinguished first by the thickness of the crust and the nature of its velocity profile, the number and nature of intermediate boundaries, and their properties.

We will be concerned basically with the possibility of obtaining an adequate depth of investigation (the upper mantle is said to extend to a depth of 1000 km; the lower boundary of the mantle is clearly recognized at a depth of 2900 km), as well as with the existence of such gross characteristics of the section as the asthenospheric wave guide of low velocity in the upper mantle.

In evaluating the feasibility of each of the methods we will base our considerations on the available data and on general concepts of a physical nature. A quantitative evaluation of the resolving power for these methods is done in Chapter III.

Surface-Wave Method. In this method, the dispersion curves for phase and group velocities are used, with a range of periods from 10 sec to 1 min for studies of the crust, and with a range from one to several minutes for studies of the mantle (wavelengths ranging from tens to hundreds of kilometers). For observations made at a single point, comparison of the observed curves for group velocity with theoretical curves for specific models permits determination of the average thickness of the crust along the travel path the waves have followed (usually some hundreds or thousands of kilometers) and the probable ratio of layers for specified two-layer crustal models.

With the study of dispersion in phase velocities using a three-station network, it is possible to determine the average thickness of the crust over a much smaller region, and to identify layers in the crust and determine their velocities by comparison of the observed curves with theoretical curves.

An investigation of the regional peculiarities in the character of the record combining all of the interference-type waves (R, L, L_g, etc.) grouped under the name of surface waves permits recognition of the type of crust present in a region — oceanic, continental, or transitional.

Thus, in the realm of regional studies, the surface-wave method may be considered as a reconnaissance method which permits recognition of anomalous crustal structure in gross regions.

Group-velocity dispersion curves have been used to determine a number of most probable models for the mantle under the continents and oceans to depths greater than 1000 km (Landisman, Aki, Dorman, Press, and others [1]).

The applicability of this method for regional studies of the mantle, and particularly its deeper parts, is not quite so obvious. It is apparent that the method allows study of horizontal inhomogeneities in the mantle for such grossly different blocks as shield regions and geosynclinal zones [201]. However, because of the interference character of these waves and their long wavelengths (hundreds of kilometers), it is difficult to expect very much resolution with this method, or the recognition of such details as changes in thickness of a wave guide or its properties within the limits of a particular block or the delineation of regions with reasonably small cross sections.

Profiles of Seismological Studies. Body waves are used in detailed seismological studies. Recording is done with high-sensitivity equipment (amplifications of 20 to 40 thousand), sensitive to frequencies from 0.1 to 10 Hz; the dominant frequencies on the records are 1-3 Hz. The equipment is sited along a specified profile, or over an area with the recording sites separated by distances ranging from 5 to 50 km. The length of a profile and the corresponding depth of investigation is unlimited.

Here we may compare the method in its profiling variant (not the areal variant) with the deep seismic sounding method, inasmuch as the observational system is fundamental to the deep seismic sounding method, also. We may recognize three basic variants to seismological profiling, distinguished by the types of waves used. In the end, though, the distinction between these variants is somewhat arbitrary inasmuch as with continuous recording of ground motion at each point caused by near and distant earthquakes, various types of waves may all be used simultaneously for interpretation. Despite this, in practice, nearly always one particular wave type is used for the primary interpretation because it gives optimum results in a particular region or under particular conditions. Thus, we may consider these variants:

1. Observations of direct waves in epicentral zones as a function of focal depth, in a manner similar to well shooting in seismic exploration (Fedotov [174-176], Lukk and Nersesov [126]).

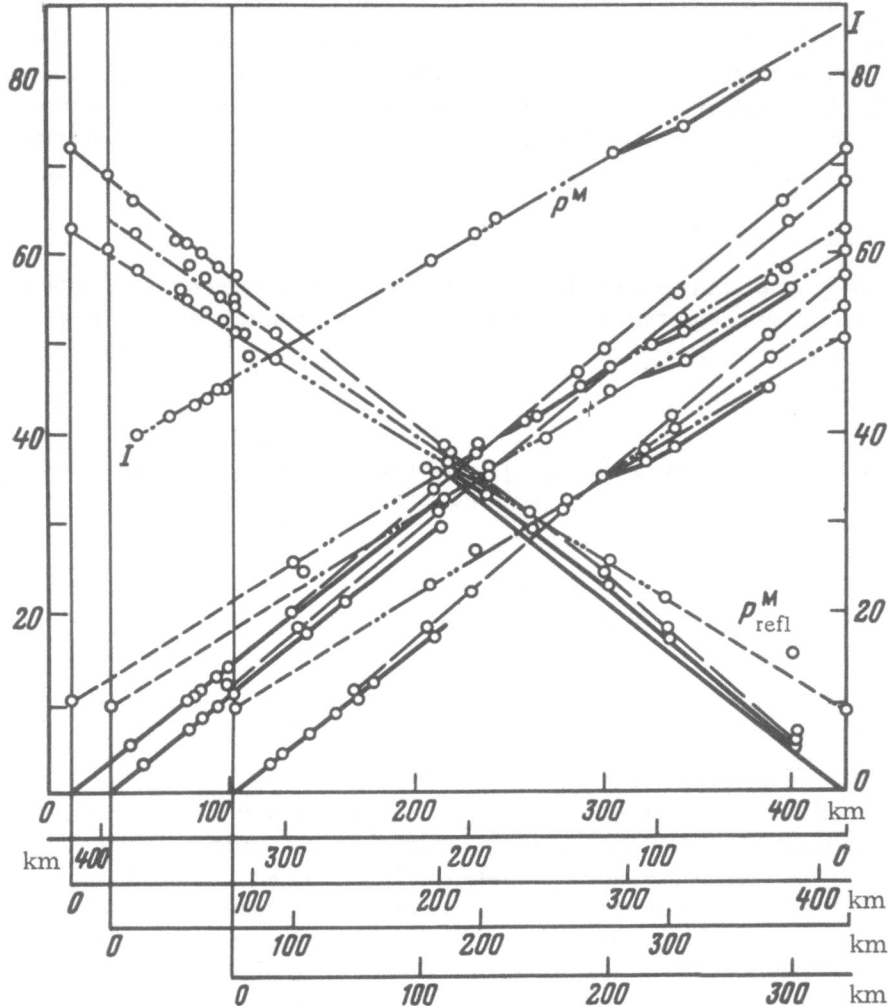

Fig. 1.1. Travel-time curves for seismological observations along in profile in the Pamir-Alai zone. The scales at the bottom are epicentral distances from earthquakes.

2. Observation of refracted and reflected waves along radial and transverse profiles located between epicentral zones, with sections being constructed from systems of reversed travel-time curves (see Fig. 1.1).

3. Observations of body waves along profiles located at considerable distance from earthquakes (up to several thousand kilometers). Usually, the best body waves are recorded in aseismic platform regions (Bulin and Sitin [29]). This approach has been widely used in the USSR since the introduction of the "Zemlya"-type equipment with magnetic recording (Mozzhenko and Pomerantseva [132]†).

Detailed seismological investigations of these three types have been used in crustal studies in the northern Tien-Shan and in the Pamir [119]. Using these data, a relief map for the M-discontinuity has been constructed for this complex region (Krestnikov and Nersesov [110]).

---

†Translator's note: "Zemlya" equipment is portable seismic recording equipment which can be left unattended at remote recording locations and records the output of an array of geophones on slow magnetic tape.

Body waves have been used also in seismological investigations in Turkmenia [29], in the Krasnodar Belt† and in Kazakhstan.‡

The precision and reliability of determining depths to the Mohorovičic boundary with observations along profiles located in epicentral zones depend on many factors. Errors which are always present are those which occur in locating the position of the epicenter and the focal depth of an earthquake. Under the best conditions an epicenter can be located with an accuracy of ±1.5 to 3 km, and the focal depth with an accuracy of ±3 to 5 km, corresponding to errors in time on the travel-time curves of ±0.5 to 0.8 sec [30, 126].

I. L. Nersesov [110] has estimated that the depth to the M-discontinuity can be determined with an average error of ±2.5 km. I consider that this is an optimistic estimate by a factor of 1.5 to 2 because of the apparent difficulty in determining focal depths accurately using travel-time curves and the effect of velocity inhomogeneities in the crust, particularly, in the upper part — in the sedimentary section.

With the frequencies observed with seismological profiling (1–3 Hz) and the rather large separation between recording sites (10 km or more), the travel-time curve approach does not make it possible to identify or trace intermediate boundaries in the crust. Neither can such information be obtained by observing arrivals as a function of focal depth because of the lack of an adequate number of earthquakes over some depth ranges and because of the imprecision in locating the source point. Furthermore, it is obvious that frequently there will be difficulties with saturation of the recording network and the recording of very weak earthquakes. Thus, at the present, travel-time curves based on detailed seismological observations may apparently be used only for studying relief of the M-discontinuity in seismically active regions.

The dilatational wave method shows some prospects for providing an answer to these problems. We can examine its utility in identifying layering in the crust, which automatically includes the problem of determining the depth to the M-discontinuity.

In this method, the difference in arrival times of longitudinal and dilatational waves formed at a common boundary is observed. The P arrival is recorded on the vertical component while the PS wave is recorded on the horizontal components. Using the difference in arrival times and knowing the velocity profile, it is possible to determine the depth to the first discontinuity at which a dilatational wave is generated, and by using the time difference for a pair of successive waves, it is possible to determine the thickness of the layer between the two successive boundaries. Hence, it follows that, being a method in which only time differences are used, the dilatational wave method is free from errors caused by uncertainty in locating epicenters, an uncertainty which is always present when travel-time curves are used. Strictly speaking, the only errors present are those related to measuring time intervals on the record and in knowing the velocities of the longitudinal and dilatational waves above the source boundary.

The resolving capacity of the method, in terms of recognizing two successive boundaries, will obviously be determined by how readily two events on the record can be resolved, and this depends on the frequency content and duration of each event.

The error in determining the thickness of any layer in the crust amounts to about 1 to 3 km. If we consider only these errors we might conclude that the resolution of this method

---

† I. V. Pomerantseva, A. N. Mozzhenko, et al. (Operations of the All-Union Research Institute for Petroleum Geophysics during the period 1964–1966).
‡ Operations of the Kazakhstan Geophysical Trust, 1965.

is only two times poorer than that obtainable with recording of vertically-incident reflections excited with shotpoints at the surface. These formal estimates of the precision of the method have been used with many crustal sections constructed using dilatational waves [29, etc.]. However, the validity of these sections, that is, their actual precision, is still largely speculative in view of our inadequate knowledge of the mechanisms required for forming a reasonably strong dilatational wave at a weak discontinuity, typical of the crust and mantle, as well as because of a number of anomalous effects which are difficult to explain for waves of this type (different arrival times for earthquakes in the same area recorded at the same point, poor correlation over short distances in many cases, and a number of other effects). A difficulty arises also in identifying the same event on records from vertical and horizontal components. All of the problems will probably be resolved in the near future.

The best dilatational wave records are observed in tectonically-simple zones — on platforms and in depressions near mountain ranges. The precision of the method is significantly less in mountainous areas.

In evaluating the applicability of the dilatational wave method, we must keep in mind that in distinction with the two preceding methods based on the use of travel-time curves, from which wave speeds may be determined, this method does not include that possibility, and therefore cannot be considered to be an independent method. It must be used in conjunction with other methods which provide information on the velocity profile in the crust. For regional surveys in a variety of zones, this condition is quite important, inasmuch as the velocity profile may change significantly.

The review presented here of the various forms of detailed seismological observations, in which waves from earthquakes are used, indicates that the method will provide a solution to problems of depth determination regionally of the following types: determination of crustal thickness and recognition of layering in the crust, under favorable circumstances, if data are available from other more detailed surveys (as for example, deep seismic sounding) to provide reference velocity profiles through the crust.

Deep Seismic Sounding. At present, frequencies of 3 to 20 Hz are used in the deep seismic sounding method. Two variations of the method are in use: the travel-time method and the point sounding method.

The travel-time method is essentially the same as refraction shooting in seismic exploration. In distinction to seismological profiling, because the source is accurately known — a shot point — the detectors can be sited exactly along a radial profile, without the uncertainty involved in knowing the spot where an earthquake will occur, and a large enough number of detectors may be placed along the profile so that a section may be drawn in as much detail as is allowed by the resolution of the method. The error involved in determining the location and the time of the excitation in deep seismic sounding is practically nil.

However, even with the deep seismic sounding method, not all of the information that might be obtained is obtained. The reasons are both logistic and economic in nature and are related to our inadequate physical representation of the true complexity of the medium being studied.

In developing travel-time curves with the deep seismic sounding method, recording arrays with various detector densities are used (see Figs. 1.2 and 1.3) and as a result, the section which is obtained is complete to a greater or lesser degree.

In point sounding with the deep seismic sounding method, a simplified version of the procedure used to generate travel-time curves is employed. In rare cases, data obtained in deep seismic sounding are inferior to data obtained with detailed seismological studies (an

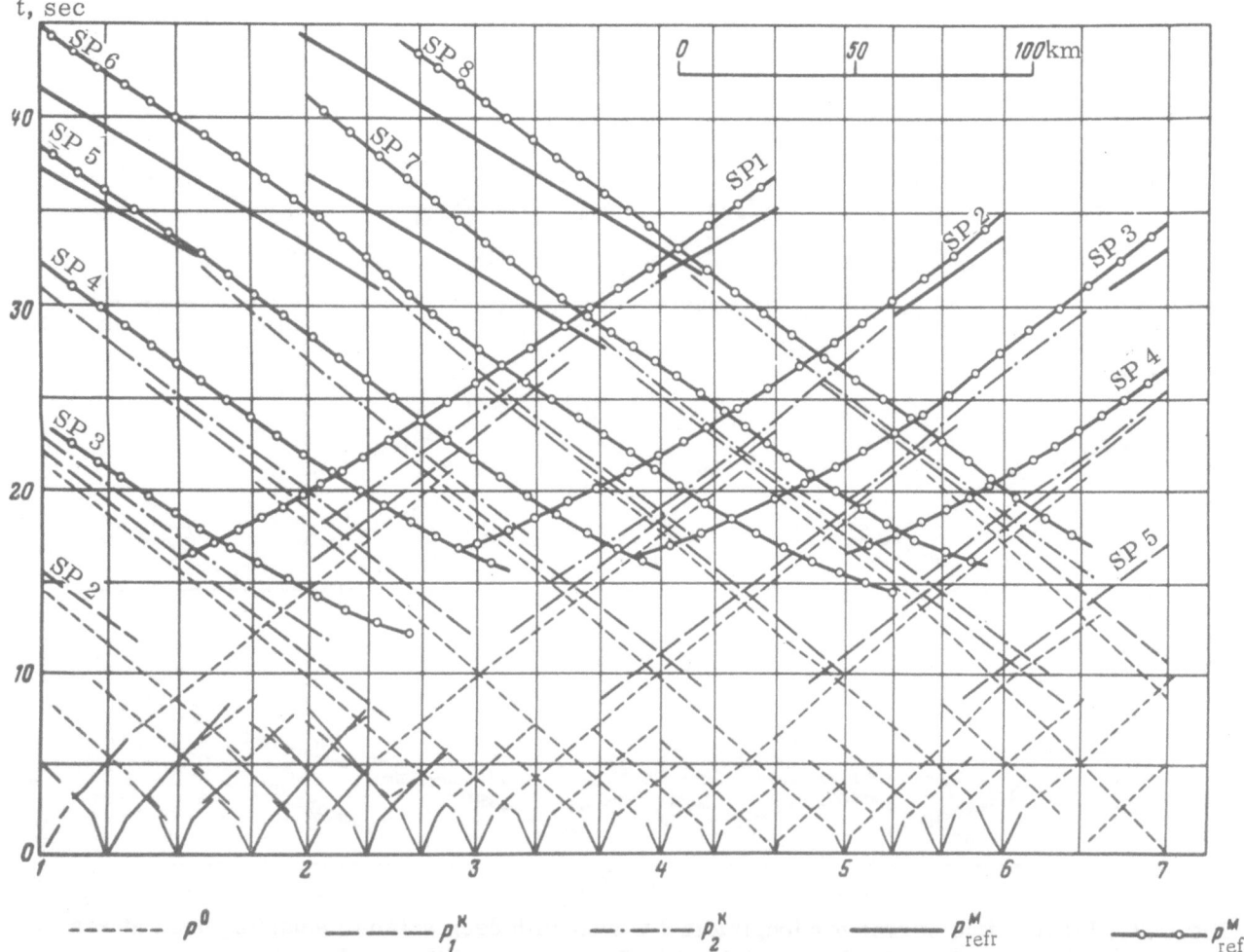

Fig. 1.2. A complete system of reversed travel-time curves obtained with the deep seismic
sounding method in the Bukhara-Khivin zone [52].

example would be an unreversed travel-time curve from a large explosion). There is also a
difference in the detail with which deep seismic soundings are carried out on land and at sea;
the detail is usually greater on land.

　　The best crustal sections are derived from continuous profiles on land using up to ten
shot points. Such data have been obtained in many areas of the USSR. Commonly, such surveys are combined with conventional reflection shooting within the sedimentary sequence, improving the reliability of the section derived from deep seismic sounding (see Fig. 1.2).

　　For exactly similar field techniques (reversed or nonreversed travel-time curves, equal
intervals between observation points, and so on) the deep seismic sounding method always provides more accurate results than seismological profiling, because the exact location of the
source is known in deep seismic sounding, and the frequencies used are higher than those for
seismological profiling by a factor of two or three. However, some auxiliary information may
be obtained with seismological profiling that is not usually obtained with deep seismic sounding.
In particular, good shear wave arrivals are recorded from earthquakes, and these are not very
well excited by the small explosions used in deep seismic sounding. These events are difficult to identify at high frequencies and are poorly recorded on the vertical component, which
is the one normally used in deep seismic sounding. Dilatational waves are essentially un-
detectable on deep seismic sounding records, while they are easily used in interpretation of
teleseismic records.

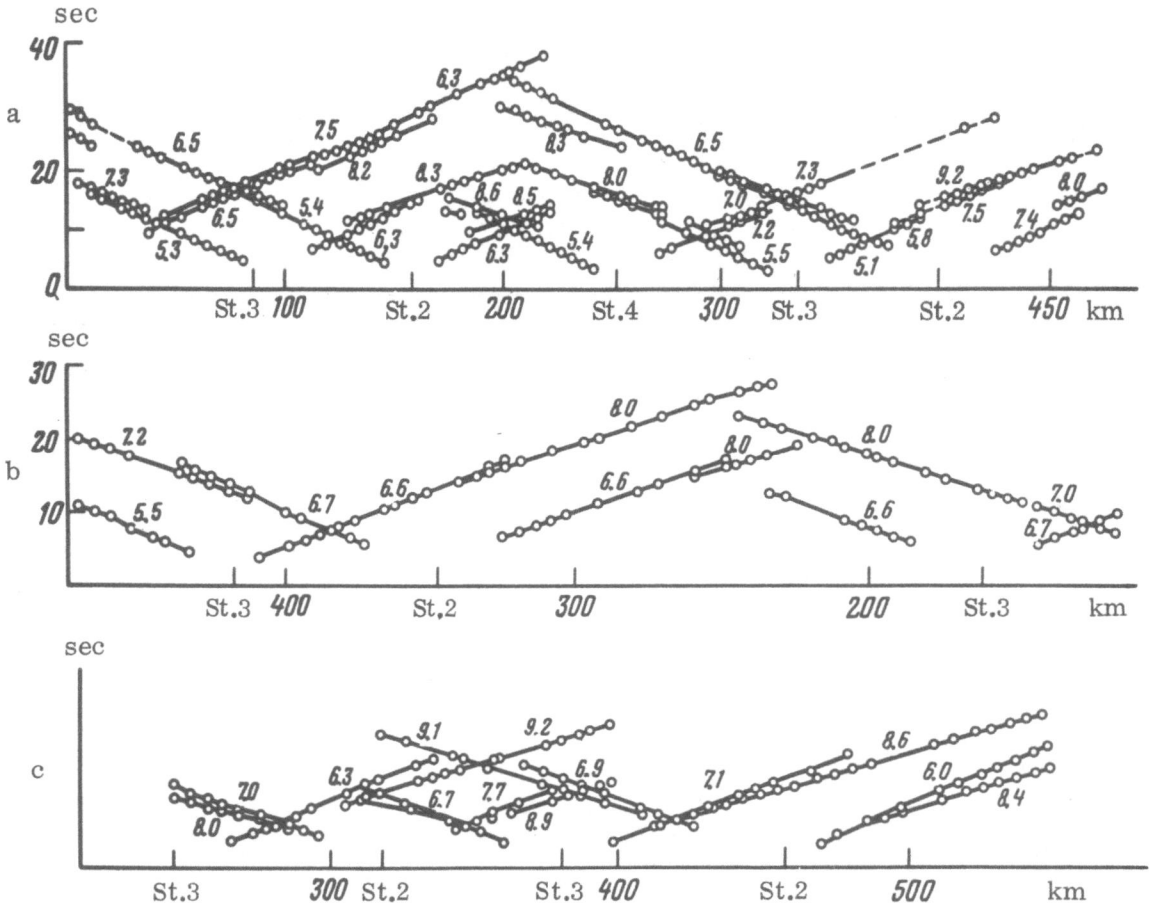

Fig. 1.3. Travel-time curves for longitudinal waves with deep seismic sounding used at sea [33]: a, b) Sea of Okhotsk; c) Pacific Ocean, east of the southern Kuriles.

It is quite important to note that with the seismological methods, there is no practical limit to the depth of investigation. By increasing the period of the recorder, it is always possible to detect the strongest earthquakes, so that readable records may be obtained for first arrivals from great distances. Thus, with seismological studies, it is possible to make observations along extremely long profiles (several thousand kilometers), so that waves which have traveled through the entire thickness of the mantle may be recorded. Additional data obtained from studies of deep earthquakes in epicentral zones have led to the construction of velocity profiles for the mantle on which waveguides for longitudinal and transverse waves are recognizable (Lukk and Nersesov, Fedotov [126, 174-176]).

The depths which may be probed with the deep seismic sounding method are presently limited to the upper mantle, not reaching to the top of the asthenospheric wave guide. This is controlled ultimately by the recording capability, which in turn is determined by the regional noise levels, inasmuch as there are basic difficulties involved in increasing the signal level by increasing the charge size. Thus, despite the apparent advantages of the various seismic methods, for regional studies of the crust and upper mantle they tend to be supplementary to one another, and there is considerable merit in using several of the methods in a single area.

The use of several methods in a single area is at present only in the conceptual stage. It is obvious that with such an approach, the techniques should be chosen to provide the widest variety of observations with the least effort. Considering the great amount of work which has

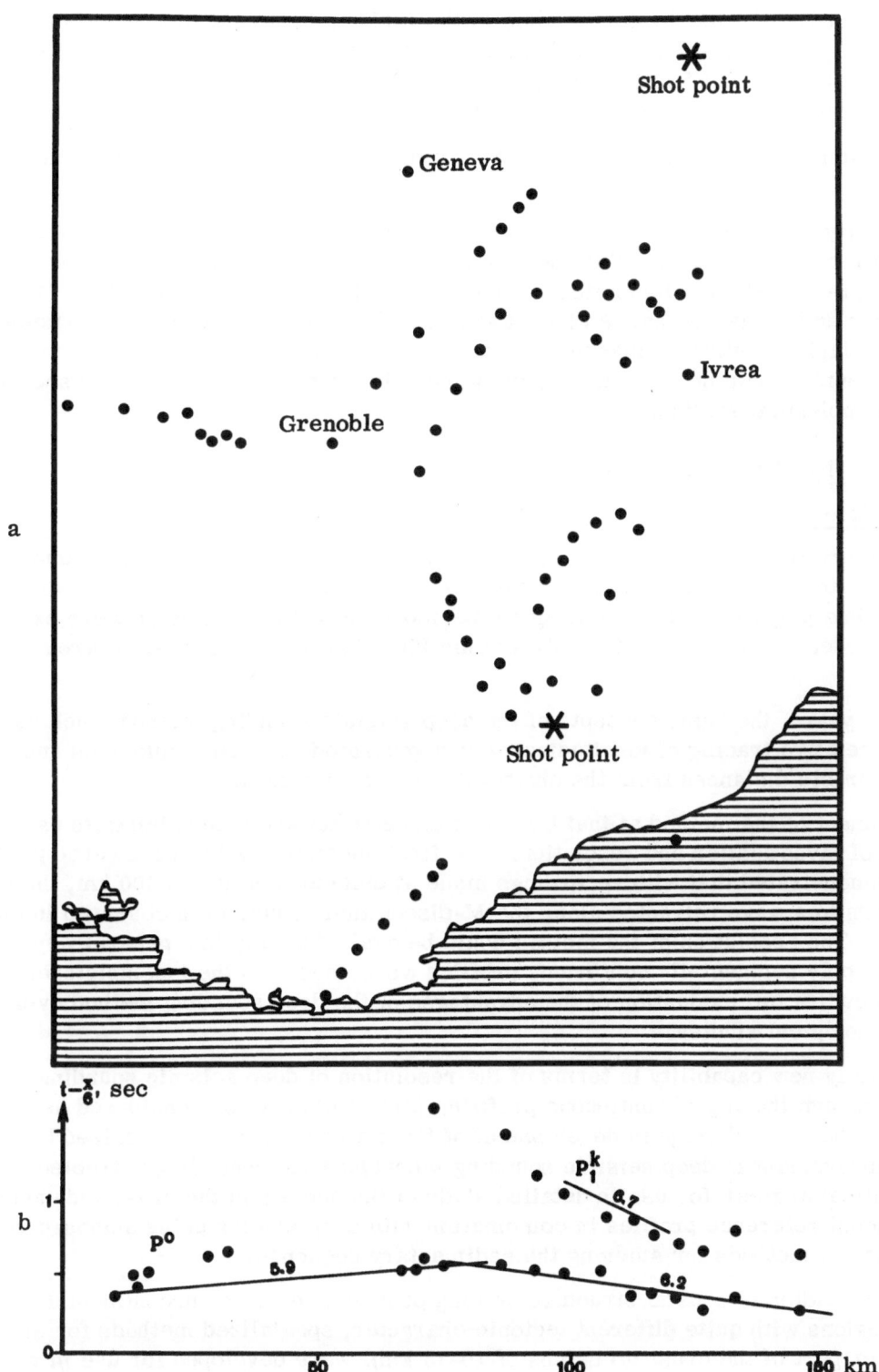

Fig. 1.4. Sparse observation network used in recording events from large explosions: a) locations of shot points and recording sites; b) travel-time curves. Work done in the western Alps during the IGY [202].

been done with regional scale geophysical studies in the USSR, a basic question in combining seismic techniques is that of developing a relationship between the highly detailed and the less detailed methods. It is apparent from a preliminary analysis that the deep seismic sounding method is the most detailed of these methods for studying the earth's crust.

The deep seismic sounding technique is the only method available for crustal studies in oceanic areas.

The practicality of the deep seismic sounding method for studies beneath the uppermost part of the mantle is not yet obvious. With the charge weights now being used and the frequency range being recorded, no information can be obtained from depths greater than 100 km on land or 20-25 km under the oceans. An increase in the probing depth of the deep seismic sounding method is highly desirable, inasmuch as detailed sections of the crust and upper mantle to greater depths could be used in improving the reliability of deeper seismic sections obtained from seismological studies.

§4. Stages in the Development of Deep

Seismic Sounding

I will review briefly the main periods in the development of the deep seismic sounding method. The first attempt at deep seismic sounding was done by G. A. Gamburtsev in the Caucasus in 1948 (see [50, 167]). Then, during the period 1949-1955, the studies were extended to northern Tien-Shan [50], western Turkmenia [50, 95] and the Pamir-Alai area [50, 96].

During these years, the basic concepts of the deep seismic sounding method, such as recording and correlative tracing of longitudinal waves generated by small explosions and detected at considerable distances from the shotpoint, were established.

The development of the method at that time was directed toward determining its capabilities — depth of investigation and resolution. The first question has been answered positively. It has been demonstrated that with recordings made at distances of 200 to 400 km, the first arrival corresponds to a wave refracted along the M-discontinuity, just as is observed in earthquake seismology. The resolution of the method could be evaluated only in a general way because at that time, only segmentally continuous profiles were used, and the incompleteness of the data did not permit a full evaluation of the advantages of the deep sounding method over conventional seismological studies.

A fundamentally new capability in terms of the resolution of deep seismic sounding was realized in 1956 through the use of continuous profiling [37]. This may be considered as the beginning point for the second stage in development of the method. It is characterized by a high degree of improvement in deep seismic sounding techniques for recording refracted waves at near-critical angles[†] for use in detailed study of the section in the crust and upper mantle along regional reference profiles in combination with deep studies using standard reflection and refraction methods for studying the sedimentary sequence.

In addition to studies of crustal structure on long profiles, nearly transcontinental in scale, crossing regions with quite different tectonic character, specialized methods for areal studies of the upper part of the crust (to depths of 10-15 km) were developed for use in re-

---

[†] Present ideas about the nature of these waves differ in that during the early development of the method, it was assumed — based on wave forms — that at large distances, all the recording waves, including critical reflections, are head waves.

gions of outcropped bedrock which had been thoroughly investigated using mining seismic techniques at mid and high frequencies (see, for example [116]).

In recent years, simplified point sounding methods have also been developed for reconnaissance of large regions [139], as well as methods for recording events from large industrial explosions using equipment with many recording channels [179].

The beginning of the second stage of development of deep seismic sounding on land coincides with the development of a marine version of deep seismic sounding in the USSR. At present, the development of deep marine surveys is proceeding in two directions: improvement of the techniques for polygonal surveys and surveys while under steam, and development of specialized methods for more detailed surveys along profiles for use in the study of continental margins, island arcs, and oceanic deeps.

During the second period in the development of deep seismic sounding, along with the improvement of techniques, an intensive analysis of the theory and methods of analysis of the observational data was begun. In so doing, some basic difficulties were encountered, caused both by the complexity of the subject and the inadequacy of the techniques available for the study. I have discussed some of these questions in earlier publications: the resolution of the method, the basis for group correlation, the nature of seismic boundaries, and so on.

Further improvements of our understanding of deep seismic sounding results is hampered by the fact that no direct control in the form of drilling is available, as there is with standard seismic surveying. Therefore, in the case of deep seismic sounding, as with many other branches of each science, a need is felt for direct penetration of the deep interior of the earth. On an international scale, this idea has resulted in the formation of several national programs for superdeep drilling, to depths of 7-15 km [18, 21, 107, 212, 213].

The first results of these programs carried out in shield areas on platforms and in the oceans will probably contribute to a new, third stage of development of deep seismic sounding. Drilling data will make it possible to evaluate the suppositions and hypotheses on the nature of the crust which serve as a basis for present methods of interpreting deep seismic sounding data.

However, it should be recognized that drilling results at specific points may not provide definitive answers to many of the more difficult problems in deep seismic sounding, and the main role of drilling in the future will be, as it now is, the confirmation of theoretical and experimental investigations carried out by other methods.

In view of the great volume of data presently available from the study of the crust with the deep seismic sounding method and with seismological methods in various degrees of detail, at the present time there is also a question about the compatibility of these data. In many cases, a comparison of results is not possible without doing further specialized work.

Considering the further development of the deep seismic sounding method for studies of the crust and upper mantle beneath the continents and oceans, it is important to make further improvements in the direction of improving the geophysical effectiveness. Up until the present, problems in the development of deep seismic sounding have been resolved in terms of obtaining a solution to structural problems, that is, in terms of obtaining a section for the earth's crust for a specified region. This has robbed the technique of the necessary completeness and generality. At present, considering that in many regions such sections have already been obtained, there are many opportunities for wider application of the method to the solution of other problems. The main one would be related to the study of seismic wave fields over a broad frequency range.

CHAPTER II

SIGNALS AND THEIR FREQUENCY CONTENT

In examining the physical basis for deep seismic sounding, it should be kept in mind that, as with any other seismic method, there are three basic sets of conditions to be considered: 1) the conditions under which the waves are generated; 2) the conditions prevalent within the medium through which the waves propagate and where they are modified to provide the information of interest; and 3) the conditions under which the waves are detected, and the selection of optimum characteristics for the recording equipment in order to provide the needed separation of signals from the noise background.

§1. Recorded Frequencies

The optimum frequency band is defined as those frequencies which provide a maximum effective sensitivity for the equipment and a maximum degree of resolution on the record for identifying and correlating individual events.

In order to specify these requirements more exactly, it is necessary to have data on the spectrum of the vibrations which are generated, the spectrum of the noise, and the properties of the medium which exert an influence on the development of the wave spectrum as they are recorded as a function of travel distance. Study of these spectrums is based on observations made with wide-band equipment and use of specialized analytical methods in the examination of records, a subject which has received intensive interest only in the last few years. During the early years of development of the deep seismic sounding method, the choice of the best frequency band for recording was done pragmatically. In studying the vibrations generated by an explosion, it was arbitrarily assumed that an explosion comprises an impulse source, generating a wide band of frequencies.

The basic consideration then devolved to the selection of the frequencies which, for small explosions, would provide the best differentiation between weak signals and microseismic noise. In this respect, microseismic noise has been treated as white noise from the point of view of its suppression statistically through the use of geophone patterns and arrays on bedrock, the selection of recording times, and so on (Gamburtsev [50]; Veitsman [35]).

Operating at frequencies of 10 to 15 Hz, using geophone patterns situated on bedrock and selecting the recording time properly, reduces the noise level and makes possible the use of a higher sensitivity in the recording equipment.

Furthermore, the upper frequency limit for recording vibrations in the deep seismic sounding method has been extended to 20-25 Hz in relation to the need to study both the crust and the sedimentary sequence with records from the same shot. On the other hand, in marine operations it has been found to be preferable to use lower frequencies, down to 2-5 Hz.

For observations that have been made on land in a number of areas under quite different shooting conditions, it has been noted that for recordings at distances greater than 100 km the predominant frequencies for deep waves lie in the range 8 to 16 Hz. Because of the lack of spectral studies, it has not been clear what the basic reason for this frequency range is: whether it is a consequence of the bulk properties of the medium or whether it may be the result of the frequency characteristics of the material immediately about the shot or close to it.

Studies of the spectrum generated by the explosions used in deep seismic sounding were first started in 1959 at the Institute of Physics of the Earth (Mikhota [121]). At that time, spectral studies of explosions, microseisms, and deep-traveling waves were done using either specially designed equipment designated as ChISS or broad band records made by the Institute of Physics of the Earth systematically in surveys on land and at sea [43, 163]. At the same time, other organizations supplied data on the systematic variations in dominant frequencies of deep-traveling waves as a function of distance for records made with narrow-band equipment [5, 42, 131, 155, etc.].

Wave spectrums have been studied over the frequency range 3-5 to 25-40 Hz on land and the range 1-2 to 10-15 Hz at sea, and the relation of these spectra to charge size, shooting conditions, and distance from the source was explored. The results of this study indicated that with various shooting conditions (in water, in wells, in dug holes), various charge sizes (from a hundred to several thousand kilograms), at various distances from the shotpoint (50 to 600 km), and under various seismological conditions (various sedimentary thicknesses, various crustal thicknesses), regular oscillations are seen on the records, with the spectra at distances greater than 50 km all being quite similar to one another (with a relative bandwidth of 30-40%) and having a peak at frequencies of 8-15 Hz on land and 4-6 Hz at sea.

Both here and in the further discussion, I will be dealing with amplitude spectra. As is the case in other branches of seismology, the phase spectrums in deep seismic sounding have only just begun to be studied.

§ 2. Shooting Efficiency

The first deep seismic sounding operations were conducted using explosions detonated in deep lakes. This procedure provided highly reproducible records and a large seismic output. When observations were made in remote areas with low microseismic noise, it was possible to record deep-traveling waves at distances up to 400 km charges of hundreds of kilograms (Gamburtsev [50]).

However, with the extension to surveys to more areas and with the requirement for detail with which recordings were made, a greater number of shotpoints had to be used so that it became impossible to use lakes exclusively as shotpoints. Shotpoints were then located in wells, dug holes, bays, openings near rivers and lakes, and so on [52-56, 152-154].

Because deep seismic sounding operations are being carried out now in many areas of the USSR where the average noise level is several times higher (sometimes as much as an order of magnitude) than that observed with the initial deep seismic sounding work in Tien-Shan, the Pamir, and Turkmenia, the average charge size used has been increased to 1-2 tons for recording distances of 200 to 300 km.

Well shooting is the form of source most widely used in land operations. Shooting is most efficient if a group of small charges (tens or hundreds of kilograms) is used rather than a single large charge. Studies have indicated that the principal factors involved in determining shooting efficiency in a shooting pattern under specific geological conditions are hole depth, the size of the charge in each hole, and the separation between holes. There is an optimum charge size for a given hole depth for which the seismic output in the desired range of frequencies (5-15 Hz) is a maximum. Thus, for a hole depth of 30 m in a region in the Ukraine

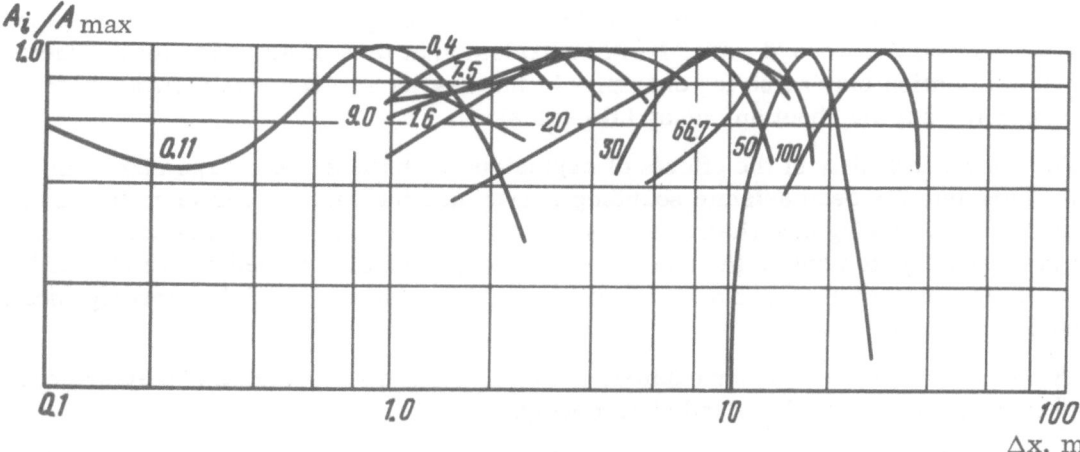

Fig. 2.1. Optimum separation between shot holes for different sizes of the individual charges (Mikhota and Tulina [123]). The figure with each curve is the charge size in kilograms.

[151], the optimum charge size is 40 kg, while under the same conditions, a charge size of 100 kg will give the same signal as 30 kg.

The best results are given by a shotpoint pattern with holes located in a triangular group. The largest amplitude seismic waves are observed for a specific separation between the shot holes which becomes greater as the charge size in each hole is made larger (see Fig. 2.1). Thus, in operations in Azerbaijan, with a charge size of 50 kg, the optimum separation was found to be about 20 m, while with a charge size of 100 kg, the optimum separation was about 30 m (Mikhota and Tulina [123]).

Because of the area occupied by such a shothole pattern, commonly it is much easier to use a linear shothole pattern, perpendicular to the direction of the profile. For a small number of shotholes (3-5) the difference between areal and linear patterns is insignificant. Linear grouping of charges along a profile is used in deep seismic sounding to suppress correlated noise, just as is done in standard seismic reflection shooting when deep reflections are to be recorded.

The medium in which a charge is situated has an important influence on the seismic effect, and this becomes particularly important for large shots. For the same charge size, different shotpoints will provide signal amplitudes on the record which differ by as much as an order of magnitude (Ryaboi [154]). However, neither in the case of standard seismic surveys nor in the case of deep seismic sounding has a reasonably reliable method for predicting the optimum shooting conditions in a medium with arbitrary properties been advanced, despite the fact that there are well-known empirical rules to the effect that shots in dry rock are less effective than shots in wet rock, shots in shale are more effective than shots in sand, and so on. As a result, optimum shooting conditions in a particular area are established experimentally.

The efficiency of shots, particularly large ones, is also determined by the nature of the explosion process itself; the packing density, the completeness of the detonation process, and so on. Experiments indicate that lack of consideration of the detonation process results in the same charge size detonated under the same conditions giving effects which differ severalfold. Therefore, in order to increase seismic signals, it is necessary to increase detonation efficiency rather than simply increasing the charge size Q, which, as is well recognized, is difficult to do in practice when large charges are being used (inasmuch as the amplitude, $A \approx Q^n$, where $n \approx 2/3$ [121]).

However, in some cases it is preferable to use the original approach, as for example when it is necessary to record waves from an explosion at tens of stations along a radial profile. Such a situation may arise for example, in the case of supplementary measurements to fill in an earlier survey about an industrial explosion.

The nonrepeatability of the effects of explosions with the same charge size complicates both the operation of a deep seismic sounding survey and the details of interpretation because of the difficulty of making quantitative use of the wave forms on the records [34]. Because of this, it is necessary to monitor the intensity of the explosions, a procedure which has been followed in most recent deep seismic sounding programs using a special recording location at the center of the profile.

The effects from underwater explosions are much more repeatable if they are carried out in deep-water basins with a fixed charge weight.

In deep seismic sounding operations carried out at sea where frequencies of 5–10 Hz are used, it is possible to increase shooting efficiency by a factor of 1.5–2.0 by selecting a charge size and detonation depth such that the vibrations generated by the explosion itself, the vibrations reflected from the sea surface, and the vibrations from the collapse of the explosion bubble are all nearly in phase [78].

With such a procedure, charges of 100–150 kg provide signals at sea which can be recorded reliably at distances of 100–150 km, while measurements may be made on shore from such explosions at maximum distances as great as 200–250 km (the microseismic noise level is usually less on land than at sea).

When several ships are used in marine operations, the charge sizes are usually the same along all profiles. If the charge size is changed, the shooting efficiency is monitored by a recording station at the shooting point, which pinpoints the shot instant. With the use of specially calibrated equipment and carefully controlled shooting conditions, this procedure is quite satisfactory.

§ 3.   Shot Spectra on Land and at Sea

It is obvious that the spectrum of waves generated by an explosion should have its peak at the frequency at which the optimum signal/noise ratio is observed in order that the shooting efficiency be maximized. Studies of shot spectra have shown that despite the fact that detonation is nearly instantaneous, in actual media a pulse structure with well-defined resonance peaks in the spectrum is formed. With larger charges, the frequency of the spectral peak is lower, but the relationship is weak. The frequency at which the spectral peak is located is essentially inversely proportional to the sixth root of the charge size (Nersesov and Nikolaev [128]). There is also a weak relationship between charge size and the shape of the spectrum. Thus, for charge sizes of 50–1000 kg, as are used in deep seismic sounding, the frequencies recorded lie in the range 5–15 Hz (see Fig. 2.2); i.e., they lie within the frequency range for which we have the optimum suppression of the noise effect.

Explosions at sea have spectra with even better developed resonance. Their form is related more closely to charge weight and depth of detonation. For a charge of 130 kg at a depth of 70–90 m, the shot spectrum has its maximum at frequencies of 3–5 Hz.

We presently judge the form of a shot spectrum from observations of the direct wave close to the shot. As the resonant character of the spectrum appears even at very short distances from the shot point [111, 239] (a few tens to a hundred meters), it may obviously be assumed that the spectrum is formed in a very small region. In this respect, experimental data agree well with theory (see, for example, Gurvich [60, 61]).

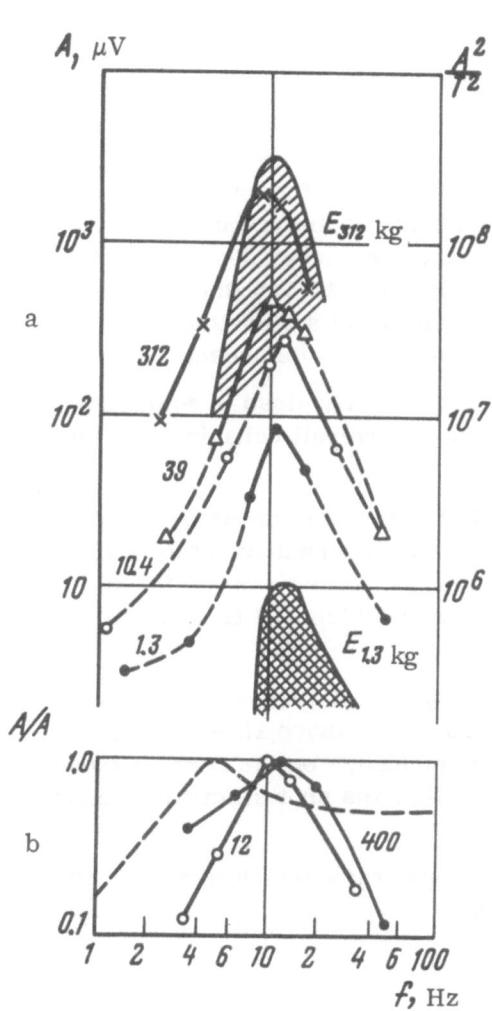

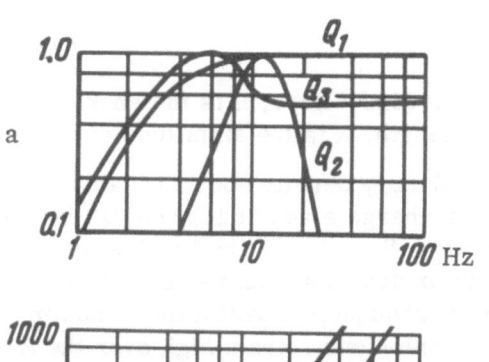

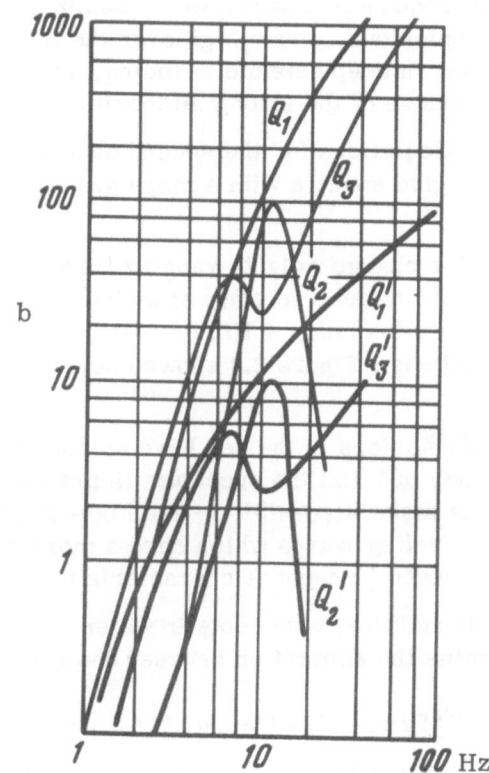

Fig. 2.2. Spectra of waves from the detonation of explosive charges of various weights (Mikhota [121, 122]; Zverev [78]): a) Detonation in wells; b) 12 and 400 kg shots in water; the dashed lines show spectra for explosions in deep sea; the values with the curves indicate charge size in kg. The shaded areas show the energy density spectra for charge size of 1.3 and 312 kg (see Fig. 2.3).

Fig. 2.3. Amplitude (a) and energy density (b) shot spectra obtained by calculations: Q_1 is a flat spectrum, land case; Q_2 is a resonant spectrum, land case; Q_3 is a sea spectrum. The energy density spectra were computed from the amplitude spectra using the formulas

$$Q_i = \frac{kA^2}{T^2}\,\Delta t \text{ with } \Delta t = \text{const,}$$
$$Q_i' = \frac{kA^2}{T} \text{ with } \Delta t \approx T .$$

So far, we have considered only the amplitude spectrum of the vibrations generated by an explosion. In considering energy, it is necessary to know the pulse length for various frequencies. We can assume either that this is the same for all frequencies or directly proportional to period. The energy will then be:

$$E = k_1 \frac{A^2}{T^2}\,\Delta t$$

in the first case, or

$$E = k_2 \frac{A^2}{T}$$

in the second. Here, E is the energy, k_1 and k_2 are coefficients of proportionality, Δt is the pulse length, A is the amplitude of a vibration, and T is the period.

Energy density spectra, calculated according to the first relationship, are shown in Fig. 2.2 for charge sizes of 1.3 and 312 kg. The maximum of the energy density spectrum is much narrower than that of the amplitude spectrum at similar frequencies. The energy density of a vibration decreases more sharply than the amplitude, particularly at low frequencies. At higher frequencies, past a local maximum and then a minimum, the energy density spectra from explosions remain high over the frequency range studied (up to 40-100 Hz) (see Fig. 2.3). However, in deep seismic sounding, the high-frequency components of a vibration cannot be used because of the strong attenuation at high frequencies at moderate distances.

Calculations for the second case with $\Delta t \approx T$, which is apparently close to actual conditions, give spectra with a more gradual left-hand slope, but the overall behavior remains the same.

For charge weights ranging from 40 to 1000 kg, most of the energy in the frequency range used in deep seismic sounding is carried by vibrations with frequencies of 8-15 Hz. At frequencies below 4 Hz, the energy density of the vibrations is lower by two to three orders of magnitude. Figure 2.3 shows energy density spectra for shots which will be referred to later.

Explosions in the sea have an energy density spectrum Q_3 with a local maximum at a frequency of 7 Hz; the spectrum is practically linear at frequencies above 10 Hz. Such explosions make it possible to use high-frequency filtering to emphasize on the records those deep-traveling waves which have a more resonant character than the shot spectrum because of the spectral response characteristics of the interfaces.

In addition, with shots in water, a strong sonic wave is generated which may be used to determine the separation between the source and receivers.

§4. Effect of Attenuation in the Crust
on the Spectrum of Deep-Traveling Waves

The spectrum of the pulse formed by an explosion changes during transmission. The main reasons for this are loss of energy in a wave because of inelastic behavior of the medium and scattering by inhomogeneities. The higher the frequency is, the stronger will be the effect of attenuation. We now have sufficient experimental data about the attenuation of the amplitude of deep-traveling waves with distance, obtained in surveys both on land and at sea, so that we may evaluate the effect of attenuation in the crust and mantle on the spectra of the various types of deep-traveling waves.

The dominant frequencies and overall spectra for deep-traveling waves observed over a wide variety of conditions may be characterized by several general aspects of their behavior [3, 42, 153, 163][†]:

1. Near a large explosion, the extent to which a dominant frequency is apparent on the records depends on the thickness of the sedimentary cover. The thicker the sedimentary

---

[†] See also: G. G. Mikhota, "Spectra of deep-traveling waves and optimum response characteristics for deep seismic sounding equipment for the conditions in western Uzbekistan," Izv. Akad. Nauk SSSR, Ser. Fiz. Zemli, No. 1 (1968).

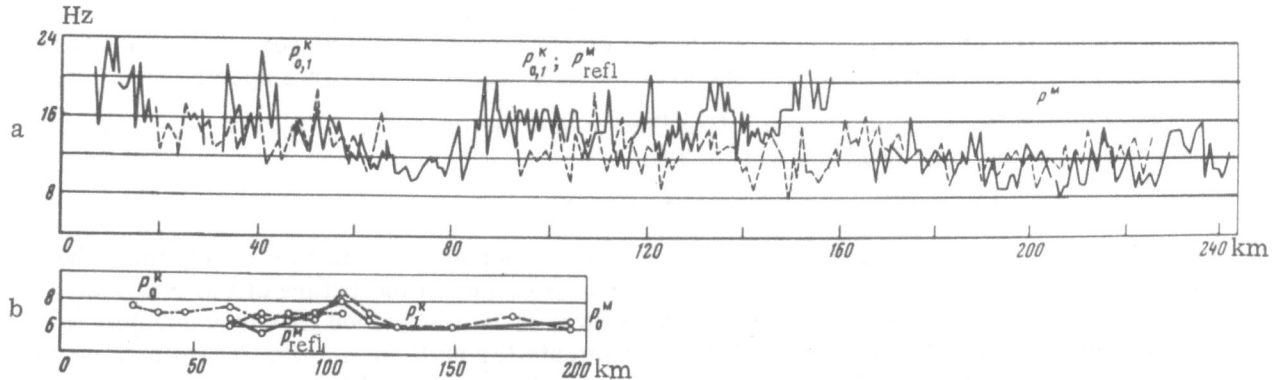

Fig. 2.4. Dependence of dominant frequency on distance from the shotpoint: a) western Turk-menia (Vol'vovskii [42]); b) central part of the Caspian Sea [2]; P_i^K and P_i^M indicate deep-traveling waves.

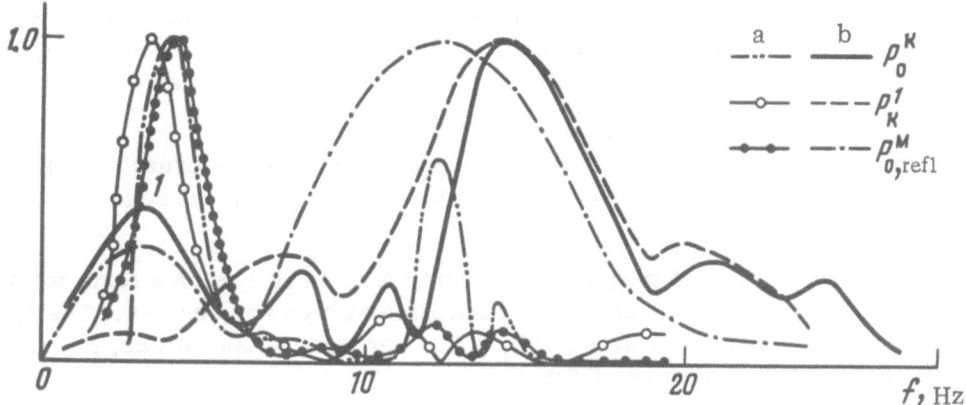

Fig. 2.5. Spectra of deep-traveling waves in relative units: a) at sea, Sea of Okhotsk [163, Chapter 7], 1958; b) on land (Mikhota, Ukraine, 1961).

cover, the lower will be the overall level of the recorded frequencies. This behavior is more complicated if there are marked changes in lithology within the sedimentary cover.

2. The dominant frequency of the oscillations on a record decreases with increasing distance from the source; at distances greater than 50 km, the rate of decrease is about 1-2 Hz per 100 km (see Fig. 2.4).

3. Usually, the spectra for different deep-traveling waves are similar (Fig. 2.5) and depend but little on distance.

However, more careful studies in a number of areas have shown that in many cases there is a small but significant difference in the spectra of several waves. That is, waves related with the surface of the crystalline basement commonly show enhanced low frequencies, while deeper near-critical reflections show higher frequencies. At large distances from the shotpoint (greater than 200 km) a significant difference develops between the frequency of the first mantle waves and later-arriving crustal events (the P_n and P^* events of the seismologists). Early arrivals are nearly always of higher frequency [36, 96].

4. In many areas, there are anomalous changes in dominant frequency, as has been substantiated through the use of various narrow-band filters.

We will consider which of these effects may be caused by attenuation. To do this, we must make an elementary evaluation of attenuation for some types of waves. In order to compare computed spectra with observed spectra, we must also compute shot spectra.

It is known from seismic prospecting theory that the spectrum of a wave $S(\omega)$ is the product of three primary functions [24, p. 13]:

$$S(\omega) = Q(\omega)\,P(\omega)\,\psi(\omega),$$

where $Q(\omega)$ is the spectrum of the waves which were generated, or the shot spectrum. The function $P(\omega)$ specifies the dependence of the wave spectrum on the nature of the interface present. Usually in studying the various types of waves it is assumed that they develop at the boundaries of thick layers, and that the coefficients for reflection and refraction do not depend on frequency, and therefore, the function $P(\omega)$ will be assumed to be a constant for each wave. The function $\psi(\omega)$ is a spectral response function which depends on the attenuation properties of the medium. It is usually expressed as $\exp[-\alpha(\omega)R]$, where $\alpha(\omega)$ is the coefficient of amplitude attenuation, and R is the travel distance for the wave.

The attenuation coefficient α, as has been shown by experimental data on the rate of decay of seismic waves observed over a wide range of frequencies and associated with boundaries within the sedimentary sequence and in the crust and mantle, is nearly directly proportional to the first power of frequency (Berzon et al. [24]).

We will compute spectra for several waves in a continental-type crust with attenuation being considered. The function $Q(\omega)$ differs for conditions on land and at sea. These are given in Fig. 2.3 and Table 1.

The spectrum $Q_1(\omega)$ is flat at frequencies greater than 10 Hz. Such a form for the spectrum is of interest for computing the effect of attenuation at high frequencies. The spectrum $Q_2(\omega)$ has a resonant character, as shown in Fig. 2.2. For marine operations, the shot spectrum has a maximum at 4.5 Hz (Zverev [78]).

Values are given in Table 2 for the attenuation coefficient at a frequency of 1 Hz; these values were selected on the basis of experimental data. If it is assumed that $\alpha = \alpha_0 f$, then at 10 Hz, a thin continental crust such as is found in coastal areas is assumed.

Spectra were computed for various wave types and distances from the source in order to explore the following aspects of the problem:

1. The effect of attenuation on the spectrum of the direct wave which is characteristic of the zone close to the source.

2. The effect of attenuation on the frequency level at the emergence points for critical ray paths of waves related to boundaries at various depths; and

3. The effect of attenuation on the spectra of reflected and refracted waves at great distances from the source point.

TABLE 1. Shot Spectra

Frequency f, Hz	Amplitude			Frequency f, Hz	Amplitude		
	on land		at sea		on land		at sea
	Q_1	Q_2	Q_3		Q_1	Q_2	Q_3
20	1.0	0.15	0.50	5	0.85	0.20	0.95
10	1.0	1.00	0.50	3	0.65	0.07	0.65
7	0.90	0.46	0.90	1	0.10	0.01	0.10

TABLE 2. Properties of the Section

Section	Layer	Velocity, km/sec	Layer thickness, km	$\alpha \cdot 1 \cdot 10^{-2}\ \mathrm{km}^{-1}$ $f = 1$ Hz
Crust	Sedimentary	2.0	1	2.5
	Upper	6.0	10	0.5-0.3
	Lower	6.6	10	0.3
Mantle	First	8.0	25	0.1
	Second	8.5	—	0.05

Spectra were computed for direct waves in the sedimentary rocks, P^S, refracted waves with travel times close to those for head waves on the surface of the crystalline crust, P_0^K, waves refracted P_1^K and nearly vertically reflected $P_{1\,\mathrm{refl}}^K$ at intermediate boundaries in the crust, waves refracted and reflected at the Moho discontinuity, P_0^M and $P_{0\,\mathrm{refl}}^M$, and waves refracted and reflected from boundaries deeper in the mantle, P_1^M and $P_{1\,\mathrm{refl}}^M$.

In computing the spectra for near-vertical reflections, we assumed that the spectrum of the reflected wave is determined by the spectrum of the incident wave $Q(\omega)$. As has been shown by the studies of N. S. Smirnova [158] and Cerveny [195-200], this approximation is valid even for the region about the source point.†

In order to simplify the computations, we assumed that the refracted waves were similar to head waves in apparent velocity. It may be shown that the spectrum of a head wave is shifted toward low frequencies with respect to the spectrum of the incident wave. According to Smirnova's computations at low frequencies, this shift amounts to 1-2 Hz. Therefore, in considering the spectra of waves shown in Fig. 2.6, only the direct and reflected waves or two head waves may be compared with each other. If a refracted wave resembles a head wave only in apparent velocity and not in wave form, then such waves may be compared directly with the direct and reflected waves.

Figure 2.6 shows spectra which take the attenuation along the ray path from source to receiver into account. A comparison of these spectra shows that in the range from 1 to 10 Hz waves corresponding to the deep interfaces have lower frequency spectra with steeper right-hand slopes. It is of interest to note that while the amount of attenuation for the P^S and P_0^M waves varies only by a factor of 1.5 in the frequency range 1-5 Hz, as may be seen from Table 3, at a frequency of 10 Hz, the variation amounts to as much as a factor of 20. At the same time the spectra of the waves $S(\omega) = \psi(\omega)Q(\omega)$ are shifted toward lower frequencies in comparison with $Q(\omega)$.

Figure 2.6a shows spectra for the direct wave P^S for various distances from the source (1 to 5 km). The right-hand slopes are shifted to the left by significant amounts, while the left-hand slopes are practically unchanged. The peak frequency on land shifts from 10 to 4 Hz, while at sea, it shifts from 6 to 4 Hz (curve 1 in this set and in the others is the computed shot spectrum for the land or sea case).

Figure 2.6b shows the spectra for deep waves in the vicinity where the critical ray path emerges. The greater the depth of the critical ray path, the lower is the spectrum of the refracted waves. Significant differences in the peak frequency and shape of the spectrum are

---

† N. S. Smirnova, using an exact formulation, has computed seismograms and spectra for waves in the region about the source point for various ratios of velocity at a boundary (from 0.6 to 0.99) and various $2H/\lambda$, where H is the depth to the boundary and λ is the wavelength in the upper layer (Fig. 4.22).

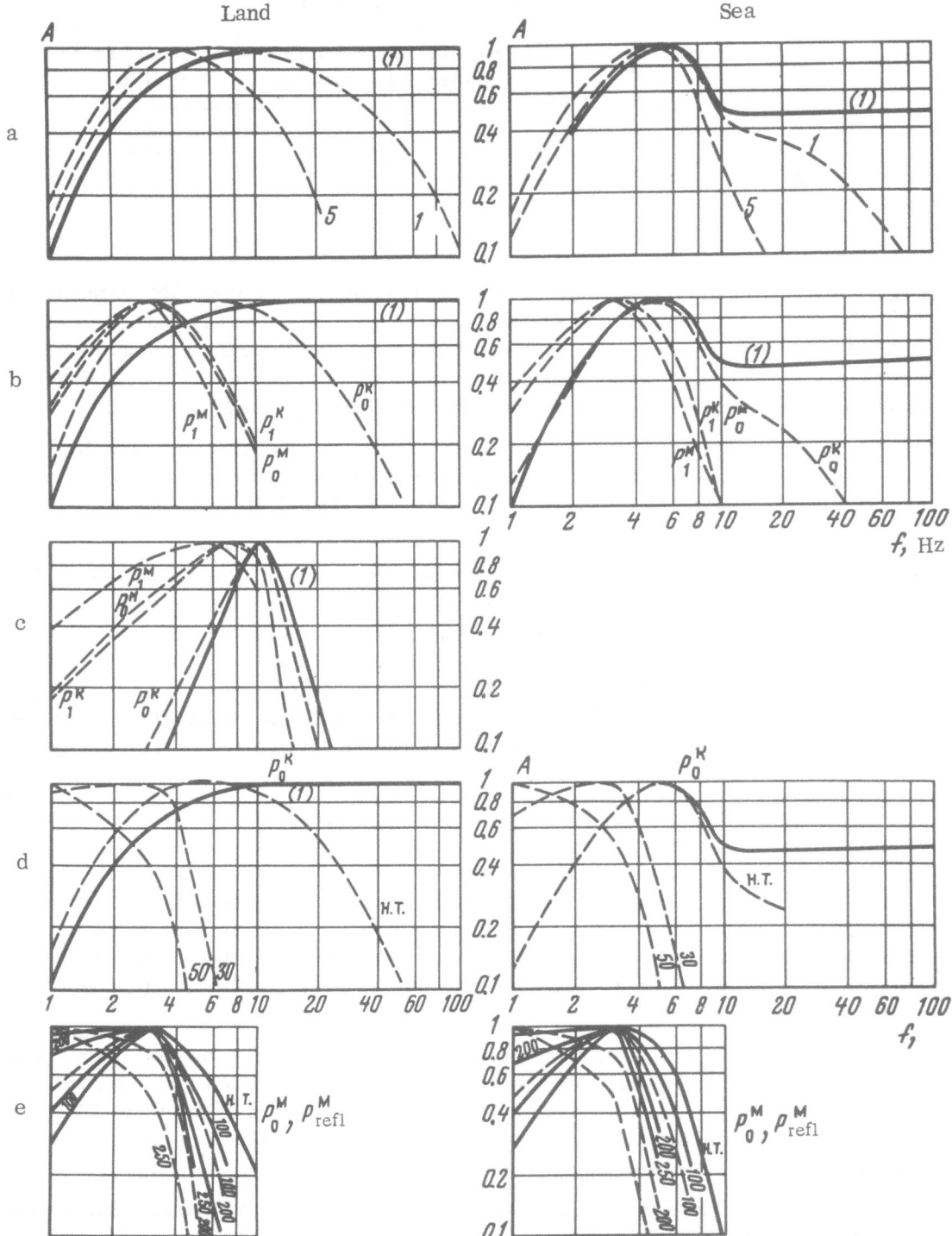

Fig. 2.6. Spectra of deep waves, computed considering attenuation in the crust. a) Spectra of the direct wave P^S for distances of 1 to 5 km from the shot point [(1) is the shot spectrum]; b) spectra of head waves at distances corresponding to critical reflection; c) the same for a resonant shot spectrum Q_2 (see Fig. 2.3); d) change of wave spectra with distance from the shot point; H.T. indicates spectra at the critical distance or at distances of 30 and 50 km; e) spectra of the waves P_0^M and P_{refl}^M (dashed) with distance in kilometers indicated on the curves.

TABLE 3. Relative Spectra of the Waves $[S(\omega)/\hat{S}](10/2\pi)$, with Consideration of
Shot and Attenuation Spectra[†]

Wave	Freq. f, Hz	ψ atten	Land Q₁	Land Q₂	Sea Q₃	Wave	Freq. f, Hz	ψ atten	Land Q₁	Land Q₂	Sea Q₃
P_0^K	20	0.347	0.59	0.084	0.58	P_0^M	20	0.0020	0.44	0.006	0.04
	10	0.589	1.0	1.0	1.0		10	0.0455	1.0	1.0	1.0
	7	0.690	1.05	0.54	2.11		7	0.115	2.27	1.16	3.72
	5	0.767	1.11	0.26	2.48		5	0.213	3.98	0.93	7.28
$R_{NT}=1$ km	3	0.854	0.94	0.10	1.88	$R_{NT}=52$ km	3	0.395	5.67	0.61	9.25
	1	0.947	0.16	0.02	0.32		1	0.783	1.61	0.20	2.64
P_0^K	10	0.00055	1.0	1.0	1.0	P_0^M	10	0.0107	1.0	1.0	1.0
	7	0.0052	8.53	4.52	16 7		7	0.0408	3.43	1.75	6.88
	5	0.0235	36.4	85.5	81.0		5	0.101	8.05	1.88	17.9
$R=30$ km	3	0.105	124	13.4	247	$R\approx 4R_{NT}=200$ km	3	0.254	15.4	1.67	30.8
	1	0.475	86.0	11.1	171		1	0.631	5.9	0.76	12.1
P_1^K	10	0.0534	1.0	1.0	1.0	P_1^M	10	0.012	1.0	1.0	1.0
	7	0.129	2.17	1.11	4.35		7	0.045	3.37	1.73	6.75
	5	0.230	2.65	0.86	8.2		5	0.108	7.67	1.81	17.2
$R_{NT}=44$ km	3	0.415	5.05	0.54	10.1	$R\approx R_{NT}=200$ km	3	0.264	14.3	1.54	28.7
	1	0.748	1.40	0.18	2.8		1	0.640	5.34	0.69	10.7
P_1^K refl	10	0.0040	1.0	1.0	1.0	P_0^M refl	10	0.00045	1.0	1.0	1.0
	7	0.0209	4.70	2.4	9.4		7	0.004	8.0	4.08	16.4
	5	0.0633	13.4	3.17	30.0		5	0.021	389	9.31	91
$R=100$ km $=2R_{NT}$	3	0.190	30.7	3.32	62	$R\approx 4R_{NT}=200$ km	3	0.098	142	15.2	290
	1	0.576	14.4	1.88	28.8		1	0.462	102	13.3	210

[†] $S(\omega)$ at a frequency f = 10 Hz is taken to be unity. The table is explained on p. 27.

observed only for P_0^K waves related to the surface of the crystalline crust. All of the deeper waves P_1^K, P_0^M, and P_1^M in the crust and mantle both on land and at sea have similar forms and are shifted toward lower frequencies in the spectrum. It is typical that the peak frequencies are essentially the same as the peak frequencies for the spectrum of the P^S wave at a distance of 5 km.

The P_0^K wave has a narrower spectrum and a lower peak frequency at sea than on land, reflecting the different characteristics of the shot spectra.

Figure 2.6c shows wave spectra for the case in which the shot spectrum $Q_2(\omega)$ has a resonant character (Table 1). In this case, the spectrum of the P_0^K wave is practically the same as the shot spectrum, while the left-hand parts of the spectra in the crust P_1^K and the mantle $P_{0,1}^M$ are shifted toward lower frequencies by a significant amount.

The peak frequencies for all waves with this form of shot spectrum $Q_2(\omega)$ are higher by 3-4 Hz than in the case of a broadband shot spectrum.

Next, spectra are shown for the P_0^K wave as a function of distance from the shot point (Fig. 2.4d); in the region of the critical angle (at a distance of about 3 km); and at distances of 30 and 50 km. A significant shift of the spectra toward lower frequencies is apparent.

In Fig. 2.6e, spectra are shown for the refracted wave P_0^M and the near-vertical reflected wave $P_{0\text{refl}}^M$ from the M-discontinuity at the critical distance (about 50 km) and at distances of 100, 200, and 250 km. The spectra of all the waves are shifted toward lower frequencies with increasing distance. In the case of the refracted waves, the peak frequency essentially does not change, but the spectra become flatter at low frequencies. The behavior is similar for the reflected waves, but the shift toward low frequencies at distances of 200 and 250 km is more marked.

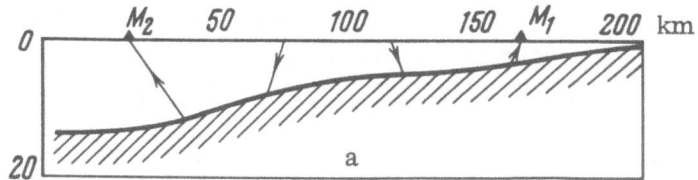

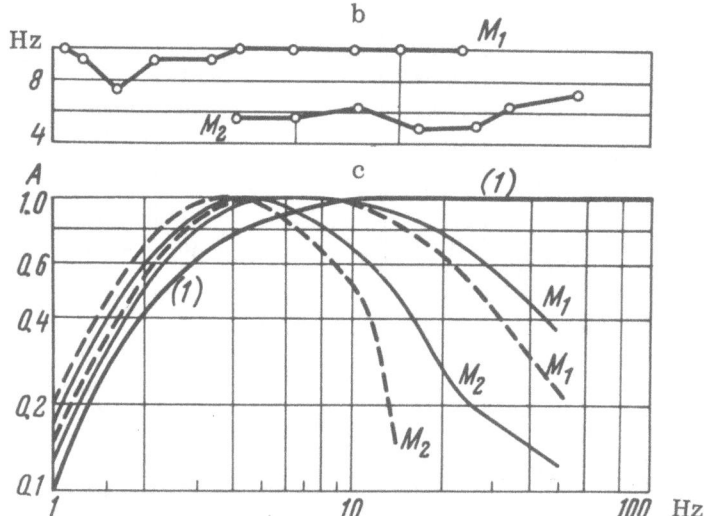

Fig. 2.7. Relation of dominant frequencies on records
to thickness of the sedimentary section: a) crustal
section along the propagation paths; b) observed curves
for dominant frequency at recording stations M_1 and
M_2; c) wave spectra at exit points M_1 and M_2 calculated
for the given shot spectrum (1). The solid curves apply
for an attenuation factor $\alpha = 3.5 \times 10^{-2}$ km in the
sedimentary section, and the dashed curves apply for
$\alpha = 5 \times 10^{-2}$ km.

Comparing these results with specific data, we have found that it is sometimes possible
to evaluate the effect of attenuation.

It is quite obvious from these calculations that the peak frequency in the spectra is
lowered significantly primarily because of the effect of the sedimentary cover and the upper
part of the crystalline crust, where the strongest attenuation takes place.

As an example of the fundamental role played by the sedimentary cover in determining
the frequency, we have computed spectra for the P_0^K wave for a specific section found in the
central part of the Caspian Sea [3, Fig. 2.7]. A comparison of the peak frequency for the computed spectra and the observed dominant frequencies indicates that the different frequencies
which were recorded at the two recording locations M_1 and M_2 are apparently explained primarily by different thicknesses of the sedimentary layers, that is, the depth to the surface of
the crystalline crust.

Therefore, the frequencies of reflected and heat waves depend on the depth of the boundary to which they are related. The deeper the boundary is, the lower will be the frequency.
The difference in frequency is most marked for the waves P_0^K, and P_1^K, P_0^M, and P_1^M. The peak

frequency of the spectrum for the P_0^K wave is about 2-4 Hz lower than for the other waves. This reflects the effect of strong attenuation in the upper layer of the crystalline crust.

Observed deep waves $P_{1,2}^K$ and $P_{0,1}^M$ usually have about the same frequency, and this is in accord with the calculations. However, experimental spectra for the P_0^K wave in many regions are even lower frequency than the spectra for deeper waves. In such cases, the spectra of the deep waves P_1^K and P_0^M are closer to the shot spectrum, while the spectrum of the P_0^K wave is shifted toward lower frequencies. This behavior of the spectra is apparently possible in actual attenuating media only if, on the one hand, attenuation plays a smaller role than we assumed earlier, and on the other, the effect of attenuation is compensated by a marked difference in the spectral response of the boundaries at which the P_0^K and the P_0^M waves are formed. If this is the case, it must be assumed that the frequency of each of the waves is determined primarily in the region where the wave is formed because the decrease in frequency with distance for all of the waves is insignificant, and essentially the same as that when attenuation is considered.

The work of S. P. Starodubrovskaya [24] has shown that thin layers possess such characteristics for reflecting waves.

If it is considered that the observed frequency for deep waves is close to the peak frequency of the shot spectrum, then in order to avoid the effect of attenuation, it is necessary to assume that the spectral response functions for deep boundaries have a dependence on frequency which is the inverse of the attenuation effect. Such spectra have been obtained for oblique angles of incidence ($i > 55°$) for reflection from thin layers ([24], Fig. 26; Mikhailova and Pariiskii [120]).

In deep seismic sounding for boundaries in the crystalline crust, the limiting angle is usually greater than 60°. As a consequence, a preliminary evaluation has shown that a strong filtering action on the spectrum of a wave may be caused by a thin layer (the program for these calculations was written by E. N. Bessonova[†]).

In order to explain the low-frequency behavior of the wave P_0^K for which there is a much smaller change in the spectrum with distance than is predicted with the assumption of attenuation, it is necessary to assume that the surface of the crystalline basement has the properties of a tuned filter with a frequency lower than the peak of the shot spectrum. In some cases, such characteristics may be exhibited by a gradual transitional layer [24, 87, 120].

This analysis of the effect of attenuation in the crust and upper mantle on wave spectra and the comparison of the calculations with observed behavior indicates that such an effect is only a partial explanation. Attenuation is the cause of the decrease in dominant frequency with increasing thickness of the sedimentary cover, and the gradual decrease in frequency with increasing distance from the shot point.

We will now examine how the relative amplitudes decrease with propagation of the various frequency components in various waves. This is directly related to the problem of what the best frequencies for recording are, with the effects of attenuation and resonance of the shot spectrum being taken into account. We will use Table 3, in which the spectrum of attenuation $\psi(\omega)$ and the wave spectrum $S(\omega)$ are given for various $Q(\omega)$ for a signal at a frequency of 10 Hz, for a quantitative comparison.

Comparison of $\psi(\omega)$ for various waves indicates that this function increases with decreasing frequency, with the rate of change becoming more marked for each wave type with

---

[†] E. N. Bessonova and G. G. Mikhota, "Interference of head waves," in: Some Direct and Inverse Problems in Seismology, Vychislitel'naya Seismologiya, Vol. 4 (1968) [see: Computational Seismology, Consultants Bureau, New York (1971)].

greater distance from the shot point. Thus, the relative importance of the lower attenuation at lower frequencies increases with distance.

Inasmuch as the shot spectrum has a resonant character, it is of interest to compare the frequencies at which we would obtain a maximal signal strength for the total wave spectrum. Such data are indicated in brackets in Table 3.

For the P_0^K and P_0^M waves at critical distances, the optimum frequency is the same as the dominant frequencies in the shot spectra, $Q_2(\omega)$ and $Q_1(\omega)$ (see Table 2). The frequency is lower for $Q_1(\omega)$; that is, for such shots, it is best to record at frequencies of 3–5 Hz despite the fact that this does not correspond to the maximum shot energy.

For distances of 100–200 km (P_0^M and P_1^M waves), the best signals are obtained at very low frequencies; 5–7 Hz on land and 3–5 Hz at sea. It is even lower for the P_{0refl}^M wave — 3 Hz on land and 1–3 Hz at sea.

Thus, at lesser distances, from 50–100 km, the effect of attenuation with a narrow-band shot spectrum is of little importance in determining the optimum frequencies. As a consequence, at these distances the best signal may be obtained by opening the response of the recording equipment to the resonant frequency of the shot, which is determined experimentally.

At distances greater than 100 km, the effect of attenuation, as shown by calculations, is no longer compensated by the resonant form of the function $Q(\omega)$ and the wave spectrum must have a maximum at a frequency lower than the resonant frequency of the shot. For the P_0^M wave, these optimum frequencies for $Q_2(\omega)$ are 5–7 Hz, while for $Q_3(\omega)$, they are 3–5 Hz. For the P_{0refl}^M wave, the frequency is 3 Hz in both cases.

Hence, it follows that at large distances, the resonant character of the shot spectrum must essentially mask the effect of attenuation for specific values of α, and so it is preferable to record the lower frequencies (3–5 Hz).

We will now determine what excitation frequencies are optimal from the point of view of the signal-to-noise ratio.

§ 5. Spectra of Microseisms and Optimum

Frequency Bands

We will now compare noise spectra with shot spectra, the attenuation spectrum, and the dominant frequencies in observed waves.

Two types of noise are recognized in deep seismic sounding. One is caused by wind, machinery, or other relatively short-term excitation. The intensity of such noise may be very high and far exceed signal levels from explosions. This type of noise is suppressed either through the use of filters or by selecting recording times during which the noise is at a minimum.

The other type of noise, which is essentially constant in level, is related to large natural noise sources, the seas and oceans. The constant component of industrial noise in heavily industrial areas should also be assigned to this class. Suppression of this noise over a large number of stations on a deep seismic sounding profile is difficult in practice.[†] As a result, the level of such noise essentially limits the amplification which may be used, or the maximum effective sensitivity.

[†] For fixed seismological recording, the regional noise may be suppressed through the use of very large arrays of seismographs, burial of seismometers in deep holes, and so on.

TABLE 4. Signal-to-Noise Ratios for Various Frequencies
(P_0^M Wave, R = 200 km, R_{crit} = 52 km)

f, Hz	$M (\omega)$	S_1	S_1/M	S_2	S_2/M	S_3	S_3/M
10	1 (1)	1	1 (1)	1	1 (1)	1	1 (1)
5	5 (3)	8.1	1.6; (2.7)	1.9	0.4; (0.6)	17.0	3.4; (6.0)
1	50 (15)	5.9	0.1; (0.4)	0.8	0.0; (0.0)	12.0	0.2; (0.8)

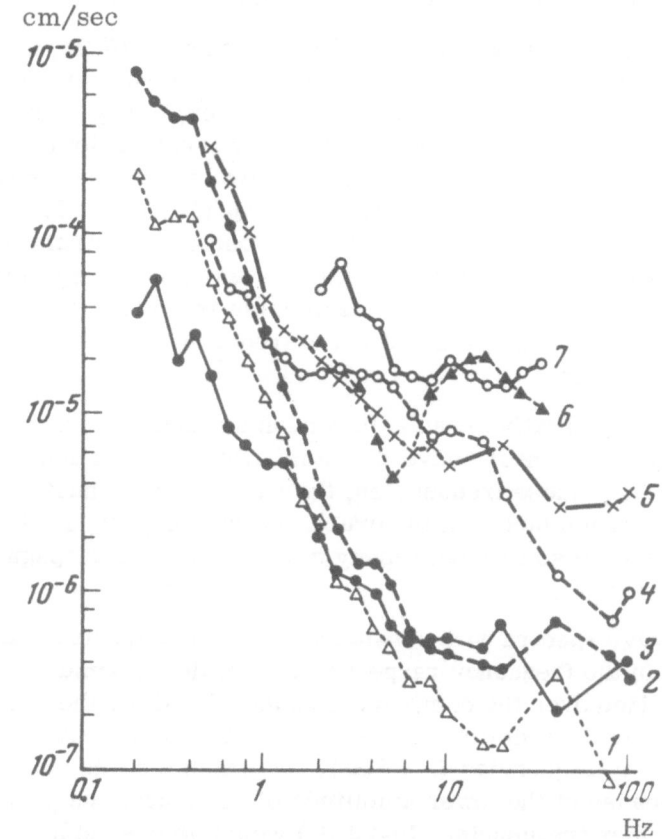

Fig. 2.8. Particle velocity spectra for microseisms
in various areas [209]: 1) Greece; 2) Puerto Rico;
3) Chile; 4) Samoa; 5) Havana; 6, 7) West Germany.

The absolute levels of regional noise differ from place to place, but the forms of the spectra are quite similar (see Fig. 2.8). For most of the world it is characteristic that the spectrum for particle velocities decreases rapidly with increasing frequency [209]. The rate of decrease is even larger for particle displacement (see Fig. 2.10).

In the frequency range for 1 to 20 Hz which is of interest to us, the minimum noise amounts to displacements of 0.1 to 400 Å, and the maximum noise levels are 70 to 10,000 Å; that is, the range of microseismic levels at a single frequency is nearly three orders of magnitude. The average level amounts to 10 Å at 10 Hz and 100 Å (10 μ) at 1 Hz.

We will compare the wave spectra $S(\omega)$ for various shot spectra $Q(\omega)$ and noise spectra $M(\omega)$ in order to obtain a quantitative evaluation of signal-to-noise levels. We will consider large distances, where the signal levels will be comparable to noise levels.

We will examine the relationships for P_0^M wave at a distance of 200 km (Table 4).

The ratio of absolute signals for noise $M(\omega)$ at various frequencies is taken from the spectra for microseisms (Fig. 2.8); the signal at a frequency of 10 Hz is taken to be unity; values of $Q(\omega)$ are taken from Table 1. In Table 4, relative intensities of the signal $S(\omega) \cdot Q(\omega)$ in comparison with signals at a frequency of 10 Hz are designated as S_1, S_2, and S_3.

The signal ratios for particle displacement spectra (see Fig. 2.10) are given in the first column, while the ratios for particle velocity spectra (Fig. 2.8) are given in parentheses. Values for the ratio S_1/M are similarly given in two columns of figures.

For resonant shots on land Q_2 the optimum frequency is 10 Hz, while at sea it is 5 Hz. Thus, the advantage of making observations at low frequencies where the effects of attenuation are not severe (the optimum frequency in Table 3 is lower, for Q_2 it is 3 to 7 Hz, and for Q_3, it is 3 Hz) does not offset the effect of increasing noise levels at lower frequencies. It is interesting to note that the optimum signal-to-noise ratio is obtained for the same frequency as the dominant frequency in the shot spectra Q_2 and Q_3 (Table 1). Thus, we must conclude that for observations in the 1 to 10 Hz band, the microseism spectrum is similar in form to the response spectrum for the attenuation of deep waves propagating through the crust. Obviously, this implies that the spectrum of microseisms recorded at any point is determined by the filter characteristics of the medium. For the frequency band 1-10 Hz, the medium apparently consists of the crust.

This can only be a qualitative result, inasmuch as microseisms, which form primarily as surface waves, need not in theory have the same attenuation as longitudinal waves. However, it is apparent that, at these frequencies, the values for attenuation are similar and are determined by the attenuation factor in the crust. In this respect, L. P. Vinnik[†] has indicated that high-frequency microseisms contain many components that propagate as longitudinal waves.

An analysis of wave spectra with consideration of shot spectra and noise spectra leads to the conclusion that in the frequency range 1 to 20 Hz, the maximum effective equipment sensitivity may be utilized over the dominant frequency band for the shot spectrum, i.e., the band from 8 to 12 Hz, which is similar to the band of observed frequencies for deep-traveling waves. The use of lower frequencies (1 to 5 Hz), for which the effects of attenuation are less, is disadvantageous because of the lower amplitude of the vibrations generated by the shot at these frequencies. Higher frequencies (15-20 Hz) would be desirable from the point of view of the output energy of the shot and the regional noise level, but such frequencies are strongly attenuated in the upper layers of the earth's crust, so that such frequencies cannot be observed in practice at the larger distances (beyond 50-100 km). Moreover, at any distance from the shot point, the local noise levels, particularly those due to wind, can be much stronger at these frequencies.

In analyzing the practicality of recording higher frequencies, it should also be kept in mind that some of these weak vibrations may be detected easily even in the presence of strong low-frequency noise levels because of a diagnostic difference in frequency. Moreover, in cases where there is a high-frequency resonant character to the reflection coefficient for a deep boundary, the high-frequency component in a record may be accentuated. This is par-

---

[†] L. P. Vinnik, Structure of Microseisms and Some Aspects of the Use of Arrays in Seismology, Izd. Nauka (1968).

ticularly significant for marine shooting, where relatively strong high-frequency energy components are generated (Veitsman [39]; Zverev and Tulina[†]).

In our considerations, we have not given specific ratios between wave spectra and noise spectra, which may vary markedly in different regions. An analysis of the detailed spectral characteristics for local and regional microseismic noise (the several minima in Fig. 2.8) may be of assistance in selecting a frequency response for the recording equipment which will permit even the recording of signals with amplitudes comparable to the noise background with an increase in signal-to-noise level of a factor of two or more for a given shot size. This is well illustrated by the work of Ivanova and Vasil'ev [83] on the optimum frequency response characteristics for studying refracted waves from the surface of the crystalline basement in several areas of the Russian platform. In studying the microseismic noise level and wave spectra, these authors showed that the maximum effective equipment sensitivity could be utilized at frequencies around 5 Hz, which corresponded to a minimum in local and regional noise and a maximum amplitude in the spectrum of the waves excited. A similar analysis of the spectra of signals and noise carried out using magnetic recording from distant shots on the ocean has shown a local minimum in the noise around 5 Hz in several areas. For shots with a resonant spectrum, signals may frequently be recorded at distances greater than 200 km from moderate-size charges (about 100 kg) (Zverev and Galkin [82]; [45, 46]).

The choice of the optimum frequency response characteristics for the recording equipment has been considerably simplified since the introduction of magnetic recording, because with broadband recording, which means 1 to 20 Hz in deep seismic sounding, this process is transferred to the laboratory, and the data may be played back through many filter settings without running the risk of losing information from a shot because of the wrong choice of filter parameters.

§ 6. Signals Recorded in Deep Seismic Sounding
and in Seismological Profiling

It is important to consider the nature of the change of signals with time for observations at a single point and the change of signals of various types as they propagate in order to develop a basis for selecting the parameters for a seismic recording channel, to provide compensation for its operation at various distances from the source, and to evaluate the possibility of using wave forms in interpretation.[‡]

In comparing signals recorded in different areas with different size charges at different distances from the shot point, it is necessary to represent them in absolute terms of motion — displacement, particle velocity or particle acceleration. This requires calibration of every channel, control of sensitivity during operation, control of amplification, and control of shot size so that observed signals can be normalized to some arbitrary shot size.

In surveys carried out by the Institute of Physics of the Earth, calibration of the amplifier equipment is always given careful consideration. Special calibration circuits, known as MGPA, have been developed for this [173]. By comparing the maximum usable amplification, which is limited by the background noise level in different areas, it has been possible to evaluate the practicality of detecting weak signals at large distances, assuming that the seismic

---

[†] S. M. Zverev, Yu. V. Tulina, et al., "Study of crustal structure in the south and central parts of the Kuriles and southern Sakhalin with the deep seismic sounding method," Report of Operations during 1963-1964, Fondi Inst. Fiz. Zemli (1966).

[‡] I. P. Kosminskaya and I. N. Galkin, "Signals recorded in deep seismic sounding," Izv. Akad. Nauk SSSR, Ser. Fiz. Zemli, No. 10 (1967).

effects from shots of similar size are essentially similar. In considering the response characteristics of the detectors — geophones or seismometers on land, pressure transducers at
sea — it has been found necessary also to evaluate the seismic signals in terms of displacement. However, because of the differences between various types of detectors and the inexact
control over their sensitivity while in use, this evaluation has been only approximate (Zaionchkovskii [76]).

Systematic determinations of the amplitude of seismic signals in terms of displacement
or pressure, with the use of carefully calibrated equipment and controlled shot size and shooting
conditions, were made in 1961 on the Black Sea, in the Caucasus in 1962 (Mikhota), in Turkmenia in 1962-1963 (Ryaboi [154]), and in the Sea of Okhotsk, the Sea of Japan, and the Pacific
Ocean in 1963-1964 (Zverev and Galkin [82]).[†]

From the point of view of analyzing wave dynamics and examining the relationship between signal size and the characteristics of crustal structure, observations at sea, where shots
are fixed in water with fixed charge size and fixed depth of immersion for both shot and detector are of the greatest interest.

In observations on land, variations in shot conditions and geophone plant conditions (the
station effect in seismology) may strongly distort the relationships caused by the structure
of the medium.

Operations with deep seismic sounding in various parts of the USSR provide the possibility of judging the relative changes in signal with time (see, for example, [163], Chapter 7), and
the relative decay of amplitude with distance from the shot point. Supplementing these data
by determinations of the absolute displacements permits evaluation of the range of signals
generated by the moderate-size shots used in deep seismic sounding comparison of these signals with observed displacements for the regional microseismic noise background in various
areas, and ultimately, evaluation of the limiting distance at which deep-traveling waves can be
detected, which in turn determines the depth of investigation.

Comparison of the signals observed in deep seismic sounding with those recorded in
earthquake seismology for earthquakes of various magnitudes permits evaluation of the capabilities of each of these methods and provides the needed background for designing an approach
based on the use of both methods.

We will consider these questions in the following paragraphs.

First, we will evaluate the signals observed in deep seismic sounding. We will consider
only the vertical component of oscillations on land and the hydrostatic pressure in water developed by longitudinal waves.

Amplitude Range for Signals Recorded at a Single Point. Modern oscillographic records
make it possible to study signals in which the amplitude varies by as much as a factor of 50
(a dynamic range of 35 dB at a single frequency). The seismic signals generated by an explosion
and recorded on land at moderate distances (up to 50 km) may show a much wider range in
amplitudes. Most of the signals observed at these distances are related to boundaries in the
sedimentary section.

If we determine the ratios of the amplitudes of waves recorded on a single seismic record
under the condition that the amplification is just high enough that the smallest detectable signal

---

[†] The idea of measuring displacements in deep seismic sounding was developed in the Group
for Combined Studies in the Department of Science of the Earth, Academy of Sciences of the
USSR. Requirements for field procedures on land were formulated in 1962 by A. S. Alekseev.

is limited by the microseismic noise background, these ratios will usually vary by no more than a factor of 10-12 (20 to 26 dB) for waves reflected from or refracted along deep boundaries. If a series of charges of different weights are used, it is possible to record signals which will be weaker than the background noise level when they are normalized for charge size [154]. In this case, the range of signal amplitudes recorded at a single point is widened in the direction of weak signals. In the following discussion we will consider the range of signal amplitudes for a standard charge size of one metric ton for land shots and of 130 kg for sea shots, sizes which are the lower limits set by noise, that is, weak signals are obtained with the recording equipment set at its maximum effective sensitivity. The dominant frequencies on land are about 10 Hz, and at sea they are 5 Hz.

For observations at sea, the range in amplitude of signals recorded with several seismic channels but at a single location using different amplifications amounts to 2 to 3 orders of magnitude. This is because waves from the sedimentary section and sound waves traveling through the water are recorded in addition to the deep-traveling waves; the sound waves are used to determine the distance from the shot point to the detector [79, 80, 163].

The difference in amplitudes of the various seismic waves decreases with increased distance from the shot point, and at distances greater than 50 km at sea or 100 km on land, the range is usually no more than tenfold.

In most areas, the following behavior of deep-traveling waves with time is observed; at distances up to 50 km or more at sea and up to 50-80 km on land, the first arrivals have the largest amplitudes, while at greater distances in continental areas, later waves have the largest amplitudes (near-critical reflections).

Change of Signal with Distance. The intensity of the first arrival decreases rapidly with distance at distances up to 50-100 km in areas with a continental-type crust, and at distances up to 50 km in oceanic areas. Over the distance range from 10 to 100 km, the first arrival usually weakens by 2 to 3 orders of magnitude, and sometimes by 3 1/2 orders, so that it drops into the background noise. At greater distances, where the first correlatable signal on land is a wave traveling through the lower crust, and at sea, a wave traveling in the upper mantle, the first arrival decreases less rapidly with increasing distance — by a factor of 5 to 10 at 100 km.

The overall change in amplitude of the first arrival for the range of distances from 10 to 200 km amounts to 3 orders of magnitude for regions with a continental crust. In the case of an oceanic crust, this same drop takes place over distances from 10 to 100 km [163].

For the later, stronger arrivals, which may have the largest amplitudes on record, signal strength decreases less rapidly with distance, with the change not amounting to more than a half order of magnitude over 100 km.

Along a single profile situated in an area where the crustal structure is uniform (without marked changes in thickness), and considering similar distances from the shot point, similar charge sizes and the same wave type, significant differences (by a factor of 2 or more) are noted in recorded amplitudes. In some cases this may be explained in terms of a relationship between a change in signal level along a profile and crustal structure. Thus, in some areas, an increase in the intensity of waves refracted along the surface of the crystalline crust in areas where this surface is buried more deeply is caused by a shift in the ray path for refracted waves, while in other cases, the amplitude of these waves decreases with increased depth to this surface because of the greater attenuation which takes place in the sedimentary section [163].

The scatter in amplitudes of mantle waves is usually less than for crustal waves within areas of uniform crustal structure [2, 163].

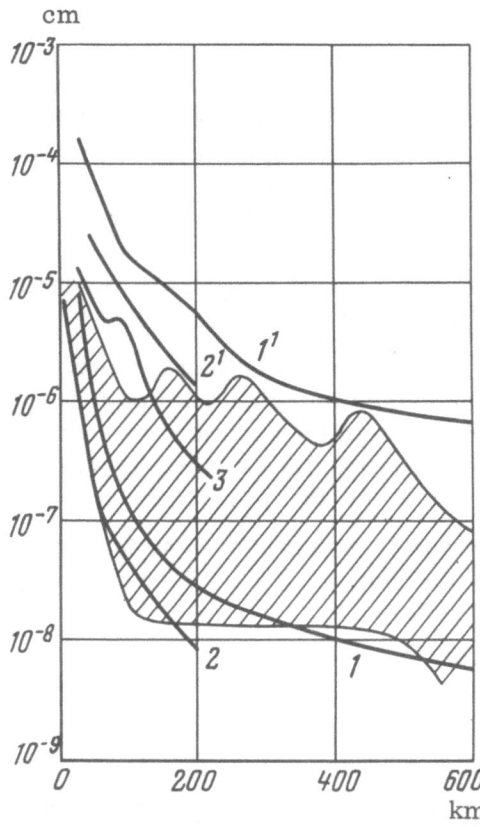

Fig. 2.9. Amplitudes of signals recorded in deep seismic sounding. The shaded area indicates signals for deep longitudinal refracted and reflected waves reduced to a charge size of one metric ton, continental crust; profile from the Aral Sea to Kopet Dag [154]: 2-2') observed signals for a charge size of 130 kg for an oceanic crust; 1-1' and 2-3) range of signal intensities for the same waves based on theoretical calculations for a uniform layered model of the crust.

      In comparing the intensities of these same types of waves in regions with a thin oceanic crust or with a thicker continental crust (shelf areas), on the average, it is found that larger signal strengths correspond to shallower depths to the M-discontinuity. This may be considered to be diagnostic of refracted-type waves [163].

      <u>Displacement Amplitude for Signals and Microseisms.</u> We will now consider the amplitudes of displacements for longitudinal refracted and reflected waves recorded on land and at sea, and compare these with the displacements for the regional microseismic noise.

      Figure 2.9 shows data on the absolute variations in displacement in deep seismic sounding. These data correspond to a charge size of one metric ton and frequencies of 10 to 15 Hz on land [154], and 130 kg and 4 to 6 Hz at sea [78]. The conversion from pressure to displacement was carried out assuming a plane wave at 5 Hz.[†]

[†] See, for example, L. M. Brekhovskikh, Waves in Layered Media, Izd. AN SSSR (1957).

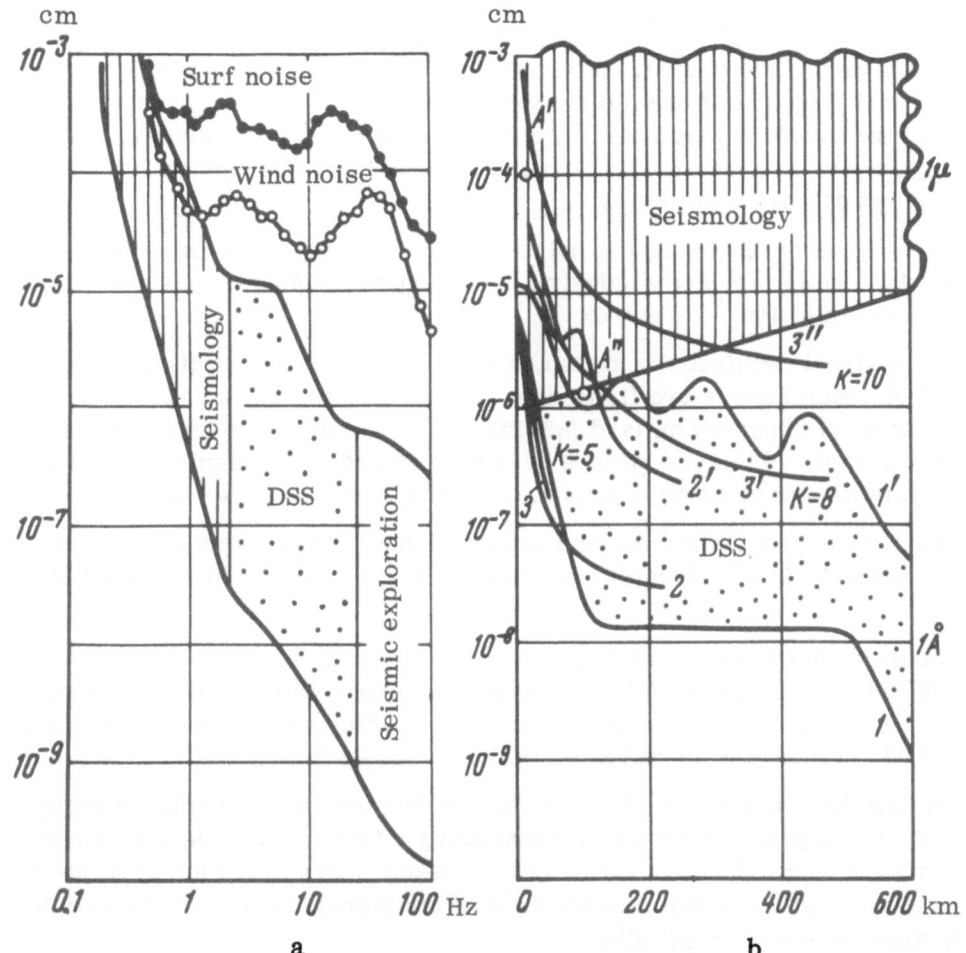

Fig. 2.10. Amplitudes of signals recorded in deep seismic sounding (DSS) and in seismology: a) displacements for microseisms, averaged data; b) deep seismic sounding signals; 1-1') on land (normalized to a charge of one metric ton); 2-2') at sea, charge weight of 130 kg; 3-3'-3") signals for longitudinal waves observed in seismology; K is an energy measure for earthquakes; A' and A" are signals from large explosions, normalized to a charge size of one metric ton.

Marine shots with a charge weight of 130 kg and a frequency of 5 Hz are essentially equivalent for the same frequency for land shots of 700–1000 kg (see § 2).

Average data on the displacements for microseisms are shown in Fig. 2.10a. These were computed from the particle velocity spectra (Fig. 2.8).

The amplitude of microseismic displacements determines the minimum useful signal which may be recorded in the presence of noise: Because the microseismic noise level varies over a wide range, we specify three grades of noise: low, moderate, and high. The useful signal must be no less than twice the level of the microseisms.

The levels of microseismic noise (in Å) are as follows:

Frequency, Hz	Low	Moderate	High
1	5	50	2500
5	1.0	10	250
10	0.5	5	50

The weakest signals of 1-2 Å, comparable to the noise level, were recognized on records from the profile Kopet Dag – Aral Sea in Turkmenia on the basis of correlation between traces [154]. The still weaker signals (0.1 Å), shown on Fig. 2.10b, merely reflect normalization of larger signals for a shot weight of one metric ton. Essentially the same order of signal (tenths of angstroms) was reported to have been recorded also by G. A. Gamburtsev in initial deep seismic sounding surveys in northern Tien-Shan and in western Turkmenia for smaller charge sizes (300-500 kg) but detonated in water.

In most other areas of the USSR, the noise level is higher, and the minimum usable signal level is about 5 Å, while in heavily industrialized regions, it may be possible to recognize only signals of the order of 100 Å or more.

The minimum signal required for satisfactory recording at sea at distances of about 200 km is about 5 Å. Such signals are limited by a noise level which is equal to the average microseismic noise level observed on land [82, 226]. For similar size shots at similar distances, the same waves may be more intense by nearly an order of magnitude, as has already been pointed out, a phenomenon which is explained by the characteristics of the medium.

The maximum signals observed on land are developed by near-vertical reflections, and their amplitudes reach 70-100 Å. In stable areas at sea, the same events have amplitudes of about 30 to 50 Å.

Thus, a standard charge weight of 1000 kg will excite signals which can be identified on records at considerable distances (200 km or more) in regions with low or moderate noise levels. In the case of a high regional noise level and with this charge weight it is not possible to record the weak first arrivals which have amplitudes lower than the microseismic levels.

Data which have been presented show that as the deep seismic sounding technique is now being used, the maximum practical recording sensitivity which is possible with ambient noise levels is nearly always used. Comparison of the frequency characteristics of deep-traveling waves and of noise also indicates that enhancement of the signal/noise ratio merely by frequency filtering would scarcely be possible.

At the same time, we see that the possibilities for recording first arrivals at considerable distance in the presence of high noise levels are quite limited. Consequently, in order to increase the depth of investigation in deep seismic sounding it is necessary either to decrease the noise (for example, by burying the geophones in drill holes or mine workings), which makes observations along a continuous profile virtually impossible, or to increase the signal level through an increase in shooting efficiency by using many small detonations in place of one large one or through the use of data processing techniques to enhance the signal strength. This last approach seems to us to provide the most promise, but it required the development of essentially new techniques.

Comparison of the Signals from Shots in Deep Seismic Sounding and from Earthquakes. We will now compare the signals recorded in deep seismic sounding with those observed in seismological studies based on a detailed regional seismograph net. The amplitudes of the signals observed at various distances from the source in deep seismic sounding and in seismological profiling are indicated on Fig. 2.10. On this illustration, the seismological curve for K = 5 and the others represent signals for P-waves excited by earthquakes of equal magnitude, K = log E, where E is the seismic energy released by an earthquake, in joules. These curves were taken from data published by T. G. Rautian [141], compiled for shallow crustal earthquakes in the Pamir – Tien-Shan region. Energy is determined from the amplitudes of the shear S-wave and the longitudinal P-waves, and in constructing the amplitude curves for P-waves as shown in Fig. 2.10, it was assumed that $A_{P+S} = 5A_P$ (Nersesov and Rautian [127]).

Data from F. F. Aptikaev [9] on displacement amplitudes in longitudinal waves from large explosions, recorded at distances of 10 and 100 km, are shown on the same graph. Signals were normalized at a charge weight of 1000 kg using the formula

$$A = Q^{0.75}$$

where A is the displacement amplitude, in cm, and Q is the charge weight, in kg. The sizes of the displacements for microseisms are indicated to the left of the illustration.

Comparison of the level of seismological signals with microseism levels at a frequency of 1 Hz with the use of amplification indicates that in areas with moderate noise levels, it may be possible to recognize signals from earthquakes having an energy measure K = 10 at distances of 500 km.

In deep seismic sounding, at a frequency of 10 Hz and at the maximum distances, amplification factors some 100 times larger than those in seismological studies are used.

Thus, at distances up to 500 km, the signals used in deep seismic sounding and in seismological studies diverge on the displacement scale; the seismological signals fall above the signals used in deep seismic sounding. There is a small region of overlap for distances in the range from 0 to 100 km, where events from weak near earthquakes having an energy production equivalent to the standard charge weight used in deep seismic sounding are recorded. These earthquakes have energy measures, K = 5 to 8. The usual minimum signal in deep seismic sounding, which is of the order of 1-5 Å, corresponds to that observed from a crustal earthquake with an energy measure, K = 4-5, at distances of 100 to 200 km, which are hardly ever used in seismological studies. The combined use of signals from explosions and earthquakes with broadband recording (1-20 Hz) may significantly broaden the applicability of seismic studies in comparison with what is possible using either deep seismic sounding or seismological profiling alone.

Recent data on signals and their spectral characteristics, as well as on the characteristics of explosions and microseisms have clarified our ideas about the nature of the medium being studied. They also confirm the splendid experimental work done many years ago by G. A. Gamburtsev on the choice of an optimum frequency band for use in deep seismic sounding, when he was advocating the development of the method.

CHAPTER III

FIELD TECHNIQUES AND RESOLUTION
IN DEEP SEISMIC SOUNDING

An examination of the question of the resolution of a seismic method, that is, the degree of detail with which the inhomogeneous velocity structure of a medium may be studied, requires the choice of a schematic model for the medium and its properties.

The detail obtainable with present-day seismic methods is determined primarily by two factors:

1. The feasibility of recognizing a specific simple wave or group of waves on a seismic record in the presence of other persistent or transient oscillations; the accuracy with which the arrival time of each wave can be determined and the number of different waves in a specified time interval are related to record resolution, primarily to the size of the contrasts and to the wavelengths; in turn, the contrasts and wavelengths depend primarily on the frequency content of the recorded signals; and

2. The reliability with which events may be correlated across records from a profile; this is determined by the characteristics of the waves and by the field procedure.

The choice of a frequency range for recording and the method of correlating events across records must depend on the properties of the medium – the shooting conditions, the propagation effects, and conditions at the detectors.

The experimental application of deep seismic sounding under a variety of seismological conditions with use of refracted waves and noncritically reflected waves, along with a great deal of work in recording deep, near-vertical reflections has resulted in the use of more general field procedures. Operational details of the field procedures used in deep seismic sounding vary. These are determined by the general form and pattern of the shot holes. Of the many possible field procedures which may be used in deep seismic sounding, we will limit our consideration here to only longitudinal profiling, because such a system reflects completely the principles involved in determining resolving capabilities of the method.

In order to evaluate resolution, we must consider data on the recording layouts which are in use and the optimum areas for recording deep events, according to travel-time curves which have been constructed for crustal sections.

In considering the various forms of profiling, only a brief review of the characteristics will be given because these techniques and their use in deep seismic sounding on land are quite similar to those used in conventional seismic exploration.

§ 1. Recording Layout

Basically, in deep seismic sounding, as in standard refraction and reflection surveys, records are made along radial profiles, with correlation between adjacent traces being used.

During the early application of deep seismic sounding, because of the limited amount of work being done and because it was being done in mountainous areas, recording sites were located at random spots distributed over an area, or located intermediately along a profile, with the use of a single shot point (Tien-Shan [50], Turkmenia [95], and Pamir [96]).

At present, three principal profiling arrangements are used in deep seismic sounding: continuous profiling, which is the method most widely used in many areas; segmental profiling, which is used primarily in mountainous areas and areas with dissected topography; and point observations, which is used on land where access is very difficult, and at sea when a moving shot point is used. In surveys near coastal areas, point measurements at sea are combined with multi-point observations on land [3, 47]. In some cases, combined profiling and areal coverage is also used with a single shot point [202, 221].

Continuous profiling is as advantageous in deep seismic sounding as it is in standard refraction surveying because it permits correlation between individual phases on adjacent records. Commonly, these two surveying methods are combined for completeness. In such cases, the distance between geophones is 100-200 m. Geophone patterns are used to average out geophone plant effects and to suppress the local microseismic noise level. A pattern usually consists of four geophones located at the corners of a square with side dimensions of 8-12 m.

Segmental profiling differs from continuous profiling only in that there are breaks in the data. The main problem with this approach arises in correlating events. This will be discussed further in Chapter IV.

Point surveys involve recording waves at a number of points which are so far apart that, in principle, correlation of individual phases on the records is not possible.

Seismological profiling is essentially a point survey method. Point surveys are used primarily for marine operations (Gal'perin and Kosminskaya [47]).

At sea, it is not possible to record while a ship is under way, as is done in standard seismic operations [79], so recording is done with the recording ship either motionless (anchored) or drifting. Meanwhile the shooting ship detonates a series of shots over a prescribed area during a specified time interval, with the effects being recorded on the recording ship. In some shipborne operations, signals are recorded from several detectors, but these are so close together that measurement must be considered as being made at a single point.

The distance between shot points ranges from 0.5 to 5 km. In surveys with one or two ships, such as are usually carried out in the more distant parts of the oceans, both the distance between shot points and the charge weight are increased as the distance to the recording ship increases. When several recording ships are used, the distances between shot points and the charge weights are held constant over the entire shooting program (3-5 km) in order to avoid confusion in operations, and this results in a decrease in the detail with which the upper part of the section can be interpreted.

Along with the development of detailed and complete correlation, surveying techniques for continuous profiling, which have become more and more detailed to meet demands for detailed interpretation of the crustal and upper mantle section, as well as the sedimentary section, in recent years, the opposite tendency may also be noted; surveys with incomplete survey techniques are coming into use. This trend is most noticeable outside of Russia, where travel-time curves are studied in the conventional manner. Our own developments are in a new direction — point sounding [137].

American geophysicists, in recording events from nuclear detonations and large industrial explosions, use a system of one-way and reversed travel-time curves with segmental correlation of phases recorded with a large number of simultaneously recording stations (up to 30-50) [214, 215, 227-229, 236].

European seismologists [203, 204, 216, 225, 230] use much data obtained with the old technique of employing recording sites distributed areally about industrial explosions, supplemented by recordings made along particular profiles and using special shots.

In West Germany, study of the deep structure of the earth's crust is done through analysis of records obtained with standard reflection exploration methods at very late times [205, 206, 222].

A method developed by Hungarian geophysicists for recording reflections from the M-discontinuity in the region near the critical angle should be mentioned as a simplified correlation survey procedure [210, 224]. Explosions and recording points are carried out along two parallel traverses in an area, with ties being made between pairs of reciprocal points. The distance between the traverses is chosen to provide optimum conditions for recording the waves. This system is supplemented by specific segmentally continuous surveys along long profiles (200-250 km).

In the USSR, simplified surveying systems based on the use of industrial explosions have not yet been sufficiently developed for use in studying the earth's crust [31, 32], but there can be no doubt about the advantages and low cost of such an approach.

Of the experiments done along these lines recently, the work of the Geophysical Institute of the Urals Branch of the Academy of Sciences of the USSR in recording mining explosions in the Urals with standard seismic exploration equipment (N. I. Khalevin et al. [179]) should be noted, as well as observations in the Altai made by the Institute of Physics of the Earth using seismological equipment (S. I. Masarskii [118]).

As a supplement to the correlation technique, which in principle should provide complete information on the velocity profile of the crust, attempts have been made to use single records to study the more legible events which may be recorded on a given seismogram.

As may be remembered from Chapter I, N. N. Puzyrev [137] has proposed a method for point sounding using refracted (or reflected) waves. This method has been developed for application in regions of Siberia where access for geophysical work is particularly difficult, primarily as a tool for studying relief of the basement surface [138]. The method has been extended to deep seismic sounding recently [139].

The point sounding method assumes recording of diagnostic strong events at individual recording locations for some optimum distance between the shot point and the recording location. In order to recognize such waves, it is necessary to use a small network of detectors about the recording location (one to two hundred meters wide). So that correlation between traces may be used in recognition of events. Soundings are located along a traverse. In order to enhance the reliability with which the characteristics of the seismic boundaries can be determined, soundings along the traverse are made alternately with long and short baselines — the distance between shot point and detector locations. Recording conditions are selected on the basis of a preliminary study of the behavior of the wave field using a radial profile as a reference.

Correlation of the data obtained from point soundings is done through consideration of all the seismic and other geophysical and geological data which are available for the region being studied. Puzyrev, who developed the method, and his colleagues have developed the techniques for the discrete correlation of waves in which they use not only the waves which can be identified individually at independent survey points, but also those which can be correlated over short distances in the survey area [139].

With respect to marine operations, note should be made of the development in recent years of interest in measurements made simultaneously on land and at sea (in the United States, West Germany, and the USSR).

In 1963-1964, during marine operations by the USSR, attempts were made to obtain detailed travel-time curves along reference traverses carried parallel to structures, in distinction to the more common practice of using reconnaissance profiles oriented across structure [81, 109]. This work also involved the introduction of magnetic recording, which improved the quality and quantity of data obtained.

In concluding this section, brief mention should be made of the problems which can be solved with the various techniques of deep seismic sounding.

Detailed Continuous Profiling makes possible delineation of the velocity of the crust and tracing of individual layers which are at least 5-7 km thick, as well as of the upper mantle, with recognition of two or three zones with thicknesses of about 10-15 km each. Using variations in velocity boundaries and anomalous seismic behavior, blocks in the earth's crust may be identified, and deep fractures may be recognized.

Supplementing the deep seismic sounding method with detailed mid- and high-frequency reflection and refraction surveys in basement outcrop areas permits study of the microstructure to a depth of 7-10 km [115, 116, 159, and others].

Low-Detail Surveys (segmentally continuous and point) which provide a reasonably complete set of travel-time curves make it possible to define the crustal section with identification of the boundary at the base of the sedimentary sequence and the one or two main discontinuities within the crust. Large crustal blocks (with dimensions of 200 km and more) may be recognized.

Point surveys at sea with a concentrated set of travel-time curves along traverses situated in quiet areas may be used to construct a crustal section with recognition of layers within the crust which have a thickness of at least 3-5 km and identification of horizontal inhomogeneities.

The simplified approach used by Hungarian geophysicists is particularly useful for determining the depth of and relief on the M-discontinuity if the basement relief is known in the survey area.

Incomplete survey systems, such as are used in the United States and in Europe, provide adequate data on the thickness of the crust and the velocity contrast at the M-discontinuity. The intermediate boundaries in the crust cannot be determined satisfactorily.

The use of long records from standard reflection surveys in West Germany has given information on the crustal section to a depth of about 30 km with recognition of 2 or 3 intermediate boundaries. However, the effective use of such data requires the application of corrections to avoid difficulties with multiple reflections which may arise within the sedimentary sequence.

§ 2. Areas for Recording the Principal Groups
of Deep-Traveling Waves

In different areas with different types of crust and using survey systems with differing degrees of detail, various numbers of waves may be recorded, with tracing of individual events

TABLE 5. Basic Wave Types

Class of wave	Velocity range, km/sec		Indices for longitudinal waves, P	
	continental	oceanic	refracted	reflected
Wave S — related to boundaries in the sedimentary sequence	Less than 5 (3-5)*	Less than 5 (1.5-3.0)	$P_{0,1}^{S}$	$P_{0,1}^{S}...\text{refl}$
Wave K — crustal waves, related to boundaries in the crystalline crust, including the surface of the basement	(5.5-7.5) (6.0-6.2)	(6.4-7.0) (6.4-6.7)	$P_{0,1}^{K}$	$P_{0,1}^{K}...\text{refl}$
Wave M — mantle events, related to boundaries in the mantle, including the Moho boundary (M_0)	Greater than 7.6 (7.8-8.2)	Greater than 7.6 (8.0-8.4)	$P_{0,1}^{M}$	$P_{0,1}^{M}...\text{refl}$

*The numbers in parentheses correspond to the usual velocities at the upper boundary of a given layer in the crust. The boundaries corresponding to the S, K, and M waves are designated in the same manner as the waves; that is, the waves $K_{0,1}...,M_{0,1}$ are from boundaries $d_{0,1}^{K}....,\ d_{0,1}^{M}...$. Dilatational waves may be designated as $S_{1,2}^{S}P_{0}^{K}S_{1,2}$ and so on.

commonly being done using the correlation approach, which will allow the tracing of deep boundaries. It should be noted that in contrast to the case in seismic prospecting, where travel-times for individual single wave forms are used, in deep seismic sounding, wave groups with similar wave speeds and wave forms are used. Various aspects of the recognition of groups and correlating them for various survey procedures are discussed in Chapter IV.

It is possible to classify the basic groups of waves commonly used in evaluating the results of a survey, using the results of the large amount of work which has been done in regions with different crustal thicknesses and different velocity contrasts. The characteristics of S, K, and M waves are listed in Table 5. Refracted and noncritical reflected waves are included in the types K and M. The conditions for identifying near-critical reflections will be discussed separately.

This table was compiled for continental and oceanic conditions from generalization of work by I. S. Vol'vovskii [42], A. V. Egorkin [69], and V. Z. Ryaboi [155] for Central Asia, by A. A. Popov for Kazakhstan, by I. V. Pomerantseva [131] for the southeastern Russian platform, by many members of the Institute of Physics of the Earth of the Academy of Sciences of the USSR for the transitional zone between Asia and the Pacific Ocean [163] and for the Caspian Sea [3], and by Yu. P. Neprochnov and others [124] for the Black Sea.

The interval over which a particular wave group in the K and M sequence is recorded as a first arrival is usually narrow, not exceeding 50 to 70 km along a radial profile. Analysis by tracing travel-time segments requires a large number of shot points. Because of the uncertainty involved with some waves related to boundaries within the crust and mantle, observations are commonly reduced to some reference event P_0^K which corresponds to the basement surface in areas with sedimentary cover, or P_0^M and $P_{0\,\text{refl}}^M$, corresponding to the lower boundary of the crust — the M-discontinuity.

For such a survey system, using the P_0^K wave as a reference, shot points must be distributed over an interval of 10-30 km to provide a travel-time curve segment with a length of 50-100 km. For the P_0^M wave, the separation between shot points amounts to 40-70 km, while the length of a travel-time segment for moderate charge sizes of 1-2 t may be as much as 250 km. With ideal shooting conditions, the length may be increased to 300-350 km. In rare cases, with this charge weight, events may be recorded at distances up to 400-600 km [152-155]. At these distances, P_1^M and P_2^M events, corresponding to boundaries within the upper mantle (up to depths of 100 km) are recorded as first arrivals.

Some of the figures also characterize the recording interval for crustal and mantle waves during marine operations. For surveys in deep water (4-6 km), the P_0^K event has a wave speed of 6.4-6.8 km/sec and is recorded over short intervals. Beginning at distances of 30-40 km, the P_0^M mantle event can be traced continuously as the first arrival, and usually can be carried to distances of 70-100 km. Events possibly related to deeper boundaries in the mantle are sometimes recorded at distances greater than 100 km. These events were first identified by Yu. V. Tulina in interpreting amplitude curves obtained along deep seismic sounding profiles southwest of the Southern Kuriles [163]. In 1964, special experimental studies were carried out to record mantle events (S. M. Zverev and others). Some theoretical analyses of the inverse problem of interpreting travel-time curves had established the possible existence of these waves (T. B. Yanovskaya, I. P. Kosminskaya, and others, 1964-1965).

Reflected waves have not been recognized in marine surveys.

A comparison of survey results with the depth of the boundaries to be studied allows an evaluation of the optimum length of a travel-time curve segment, in terms of the depth of a boundary. This length amounts to 10-20 times the depth for the surface of the crystalline basement, and 5-7 times the depth for the M-discontinuity and boundaries in the upper mantle. In initial operations in a new area, these figures may be used as a means for predicting the necessary recording parameters.

We may conclude on the basis of the data presented here for the recording ranges for the principal K and M events that at present we have adequate information for specifying the primary factors for the correlation survey system to study the relief and velocity contrast of boundaries deep in the earth's crust and upper mantle to depths of about 100 km under the continents.

Deep seismic soundings at sea have been limited to studies of the earth's crust at depths of 15-20 km, although greater depths have been reached in specific cases [109].

§3.   Optimum Ranges for Recording Deep Reflections

The problem of recording near-vertical or low-angle reflections in deep seismic sounding differs for each layer.

With respect to the use of near-vertical reflections, methods based on their recognition are still experimental, despite the fact that a great amount of work has been done. The reason is that at the distances at which near-vertical reflections from deep boundaries are recorded under the continents, which are 30 to 50 km from the shot point, there are difficulties involved in correlation. The distance over which such events can be correlated in deep seismic sounding is only a few kilometers, that is, a small fraction of the depth to the reflecting boundary.

Moreover, when the reflection is recorded close to the shot point, the amplitude of the reflected event is usually comparable with that of the background noise level. Theoretical considerations have also indicated that deep reflections at near-vertical angles must be weaker than shot-generated noise associated with other types of waves [5, 163]. This is clearly evident from Fig. 3.1, which shows the amplitudes of reflected waves and background noise along with theoretical curves for the amplitude of reflected and head waves.

Close to the critical angle, beginning essentially at 20-30 km from the initial point of the travel-time curve for the refracted waves, the character of reflections in nearly all areas

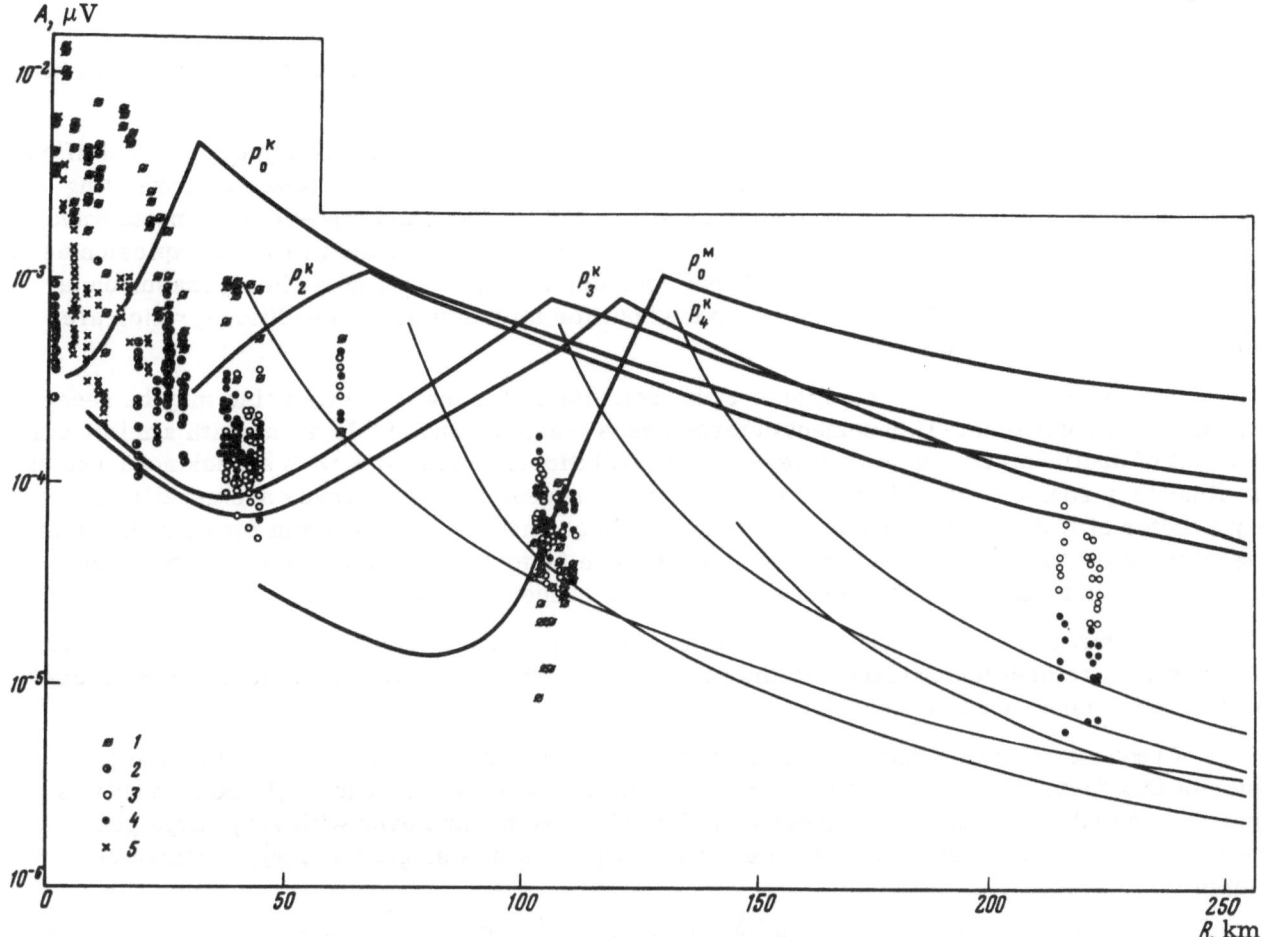

Fig. 3.1. Amplitude characteristics of deep waves (Kazakhstan, Karazhal deep seismic sounding profile, 1962). Observed amplitudes are given in microvolts for a standard charge weight of 100 kg with normalization using the formula $A = Q^n$, where $n = 0.5$ for $Q < 100$ kg and $n = 0.75$ for $Q > 100$ kg and with narrow band filtering at 10 Hz. 1) Refracted waves as the first arrival; 2) reflected waves at late times (later than 10 sec) at distances of less than 50 km from the shot point; 3) reflected waves from intermediate boundaries in the crust; 4) reflected waves from the M-discontinuity; 5) nonstationary seismic noise observed in the time interval during which near-vertical reflections from boundaries in the crust arrive. The curves show the theoretically predicted amplitudes of reflected and refracted waves for crustal sections similar to those constructed from deep seismic sounding travel-time curves. The theoretical curves are compatible with the P_0^K curves compiled by G. A. Yarshevskii using seismological data [188] and data from the Institute of Geology and Geophysics of the Kazakh SSR.

improves, becoming relatively more intense and in most cases, being continuously traceable, both in the area about the critical point and beyond the critical point, up to distances of 30 to 100 km from this point. This is the optimum region for recording deep reflections.

Such an area has also been found in many regions for the P_{1refl}^K event corresponding to the lower crust, as well as for the P_{0refl}^M event, corresponding to reflections from the M-discontinuity. The interval for optimum recording of the P_{0refl}^M waves in shield and platform areas is 70 to 200 km, that is, at distances of 2 to 5 times the depth [113, 114, 129-131, 160-161]. The P_{0refl}^M event is usually dominant on records made over this range of distances from the shot point.

For crustal waves related to boundaries in the upper crust, the range for recording clear events at less than critical angles is usually less, while for more than critical angles, it is not possible to trace such events because of interference with other waves.

Wave Noise. We will now consider the principal wave noise which hampers the identification of deep reflections at less than critical angles. This noise may be viewed as consisting of low-velocity noise waves (wave speeds less than 4-5 km/sec) and high-velocity noise waves (6 km/sec and more), and as nonstationary noise — vibrations associated with the explosion used to generate the seismic signals. We will not consider microseismic noise here, inasmuch as in practice, a deep seismic sounding is carried out for minimum noise conditions, which are lower than seismic noise.

Two types of areas having different characteristics for low-velocity noise may be recognized on the continents: 1) basement outcrop areas — shields and stable areas with a minimum amount of sedimentary cover; 2) covered areas — platforms and basins with a thick sequence of sedimentary rocks (several kilometers). The wave speeds for these waves are usually lower by a factor of two than the wave speed for longitudinal waves. The spectrum for wave noise is usually biased toward low frequencies. All of these factors contribute to the feasibility of using frequency filtering and geophone patterns for reducing noise.

In covered areas, the low-velocity noise is contributed primarily by longitudinal and dilatational waves related to boundaries in the sedimentary section, as well as by multiply refracted waves in the upper part of the crust.

Frequency filtering and the use of geophone patterns usually provides excellent suppression of this form of noise, but in some regions, as for example, in Bukhara,[†] the intensity is nearly two orders of magnitude higher than that of the signal and even with very large geophone patterns (for example, with the use of 20 geophones in a single pattern), it cannot be suppressed adequately.

High-velocity stationary noise is characteristic of both types of areas. It is generated by various wave groups arising at boundaries in a layered crust, including various multiply reflected waves.

Some effort has been made to study this type of noise at various distances from the shot point using the directionally controlled reception method (Ryaboi and Vol'vovskii [20, 42]). The clearest results were obtained in the range for optimum recording of reflected waves in passing through the critical angle. In this range, deep reflections commonly dominate on the records, and so, may easily be distinguished from noise.

There has been virtually no study of high-velocity noise close to the shot point. The difficulty with such a study is that the intensity of nonstationary seismic noise is comparable.

Considering results obtained in routine seismic exploration, it can be assumed that when several boundaries are present in the earth's crust, multiply reflected waves will be developed which make it difficult to recognize singly reflected events.

We will now evaluate which of the multiply reflected waves may be most deleterious in terms of masking the correlation of deep reflections from primary boundaries.

In covered areas, multiply reflected events may arise at boundaries within the sedimentary sequence. However, even with a high degree of multiplicity, such events will hamper

---

[†] N. I. Davydova, A. N. Fursov, and G. A. Yaroshevskaya, "Report of attempts to record reflections at less than critical angles with PMZ equipment in the Bukhara region during 1964-1965," Inst. Fiz. Zemli, Akad. Nauk SSSR, Uzbek. Geophysical Trust, Moscow (1966).

TABLE 6. Reflection Coefficients for Singly Reflected Waves from Boundaries in the Crust and Mantle for Vertical Incidence and at the Critical Angle

Type of crust	Layer number	Thickness, km	Density, g/cm³	Velocity, km/sec	Velocity ratio at boundary bet. layers	Reflection coefficient for vertical incidence	Reflection coefficient at the critical angle
Continental	1	10	2:50	4.3			
	2	8	2.75	5.9	0.729	0.200	0.728
	3	16	2.90	6.4	0.920	0.069	0.990
	4	16	3.05	7.4	0.865	0.097	0.960
	5	8	3.05	8.4	0.880	0.110	0.980
	6		3.35	8.6	0.980	0.012	0.990
Oceanic I	1	5	1.0	1.5			
	2	2	2.0	2.0	0.750	0.455	~1.000
	3	5	2.8	6.8	0.294	0.652	0.716
	4		3.3	8.2	0.830	0.173	0.820
Oceanic II	1	5	1.0	1.5			
	2	5	2.8	6.8	0.221	0.855	~1.000
	3		3.3	8.2	0.830	0.173	0.820

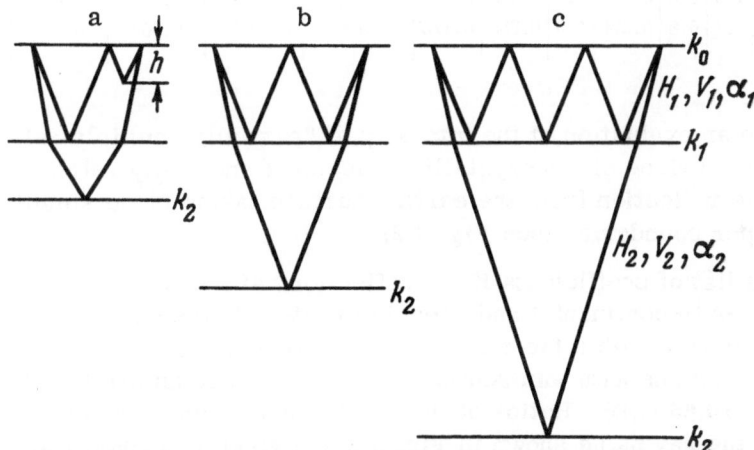

Fig. 3.2. Ray paths for singly reflected and multiply reflected waves [72]: a) partially reflected; b) double multiple; c) triple multiple.

the recognition of primary reflections only for boundaries that are located relatively high in the crystalline crust, at depths up to 5-10 km. For a sedimentary sequence with a thickness of several kilometers, as indicated in reference [222], even the fifth multiple is limited to a duration of about 3-5 sec. This computation is based on a frequency of 20-30 Hz.

Because well-defined sedimentary boundaries usually correspond to thin layers, it may be assumed that for the low frequencies used in deep seismic sounding (~10 Hz), these boundaries will appear less sharp and play a smaller role in the formation of a partially multiple reflected wave from a deep boundary than, for example, the surface of the crystalline basement. Therefore, we will not consider multiple reflections within the sedimentary section further.

TABLE 7. Partially Reflected Waves

Type of crust	h_{pr}, km	H_1, km	H_2, km	V_{pr}, km/sec	V_1, km/sec	V_2, km/sec	$p = \frac{V_1}{V_2}$	K_0	Normal incidence						Critical angle				
									K_{pr}	K_1	K_2	$\alpha_1 \cdot 10^3$, km^{-1}	$\alpha_2 \cdot 10^3$, km^{-1}	$\frac{A_{single}}{A_{multiple}}$	K'_{pr}	K'_1	K'_2	$\frac{A_{single}}{A_{multiple}}$	x_{crit}, km
Continental	4	18	16	2.5	5.2	6.4	0.816	1.0	0.3	0.069	0.097	7.5	4.0	5,20	0.16	0.87	0.92	1,93	80
	4	34	16	2.5	5.55	7.4	0.750	1.0	0.3	0.097	0.110	7.5	3.5	4.4	0.15	0.8	0.98	3,5	97
	4	50	8	2.5	5.9	8.4	0.700	1.0	0.3	0.110	0.120	7.5	2.6	0.43	0.25	0.27	0.15	2,68	20
Oceanic	2	7	5	2.0	1.5	6.8	0,221	1.0	0.652	0.652	0,173	5	3.0	0.04	0.716	0.716	0.820	0,135	—

TABLE 8. Fully Reflected Waves

Type of crust	H_1, km	H_2, km	V_1, km/sec	V_2, km/sec	$p = \frac{V_1}{V_2}$	K_0	K_1	K_2	α_1, km^{-1}	α_2, km^{-1}	$\frac{A_{single}}{A_{double}}$	$\frac{A_{single}}{A_{triple}}$	Critical angle			
													K'_1	K'_2	$\frac{A_{single}}{A_{double}}$	$\frac{A_{single}}{A_{triple}}$
Continental	18	16	5.2	6.4	0.812	1.0	0.069	0.097	$7.5 \cdot 10^{-2}$	$4 \cdot 10^{-2}$	40	1360	0.5	0.960	5.8	2.7
	18	32	5.2	6.9	0.754	1.0	0.069	0.110	$7.5 \cdot 10^{-2}$	$3.5 \cdot 10^{-2}$	45	1720	0.8	0.980	1.1	15.1
	18	40	5.2	7.4	0.703	1.0	0.069	0.012	$7.5 \cdot 10^{-2}$	$2.6 \cdot 10^{-2}$	6.5	340	0.78	0.990	2.2	9.4
Oceanic	5	2	1.5	2.0	0.750	1.0	0.455	0.652	$5 \cdot 10^{-4}$	$1 \cdot 10^{-1}$	0.465	0.250	0.290	0.716	1.4	1.2
	5	5	1.5	6.8	0.221	1.0	0.855	0,173	$5 \cdot 10^{-4}$	$3 \cdot 10^{-2}$	0.002	0.00	1.0	0.820	0.00	0.00

We will give an evaluation of the intensity of "partially-multiply reflected" reflections associated with the surface of the crystalline basement and "fully reflected" waves in the same crust with multiple reflection from the earth's surface taking place simultaneously with reflection from deeper boundaries (see Fig. 3.2).

Table 6 is a list of coefficients K for reflections at normal incidence and at the critical angle for boundaries in continental and oceanic crusts. At the surface of the crystalline crust, K = 0.2. For boundaries within the crust, it is less than 0.15, and at boundaries in the mantle, less than 0.05. In regions with thin sedimentary layers such as are found with an oceanic crust, K amounts to 0.65-0.86. Ratios of the amplitudes of singly reflected and multiply reflected waves for the ray paths shown in Fig. 3.2 are given in Tables 7 and 8.

This system has been analyzed in a paper by A. M. Epinat'eva [72] for the case of conventional seismic exploration. Equations are given in that reference which assume the following form for partially reflected waves:

$$\frac{A_{single}}{A_{multiple}} = e^{2h\,\alpha_1\left(1 - \frac{\alpha_2}{\alpha_1}\frac{1}{p}\right)} \frac{1 + (h/H_1)}{1 + \frac{h}{H_1}\frac{1}{p^2}} \frac{K_2\left(1 - K_1^2\right)}{K_0 K_1 K_{pr}}$$

while for fully reflected waves,

$$\frac{A_{single}}{A_{multiple}} = e^{2H_1\alpha_1\left(1 - \frac{\alpha_2}{\alpha_1}\frac{1}{p}\right)} \frac{2}{1 + (1/p^2)} \frac{K_2\left(1 - K_1^2\right)}{K_0 K_1^2},$$

$$\frac{A_{single}}{A_{multiple}} = e^{4H_1\alpha_1\left(1 - \frac{\alpha_2}{\alpha_1}\frac{1}{p}\right)} \frac{3}{1 + (2/p^2)} \frac{K_2\left(1 - K_1^2\right)}{K_0 K_1^3}.$$

The symbols used in these expressions are defined in Fig. 3.2; α_1 and α_2 are attenuation factors, $p = V_1/V_2$ is the ratio of velocities, K_{pr}, K_1, and K_2 are reflection coefficients at the intermediate, and first and second boundaries, and h_{pr}, H_1, and H_2 are the thicknesses of the individual layers.

In our computations, we have represented a many-layered medium as being three-layered. The boundaries with reflection coefficients K_1 and K_2 are located at depth, while the boundary with reflection coefficient K_{pr} is undefined. In case a, Fig. 3.2, the velocity V in layer H_1 is computed from the formula for average velocity. The values for the factors α_1 and α_2 do not change.

For a continental crust, equal travel times for singly reflected and multiply reflected waves (for the boundaries between layers 4, 5, and 6 in Table 6) will be observed at rather large distances, about 50 to 100 km. Therefore, in computing the intensities of these waves, it is necessary to consider the change in reflection coefficient with increasing angle of incidence. Values for K_1' and K_2' computed using tables by Petrashenya and others [43] are listed in Tables 7 and 8.

The significance of these corrections may be judged from Table 6, in which reflection coefficients for both vertical incidence and critical angle are listed.

It is obvious from the ratios of intensities (Tables 7 and 8) that partially reflected waves related to a well-defined basement surface may cause some considerable difficulty in recognizing reflections from boundaries in the upper part of the crust (Table 6, boundary between layers 2 and 3). There is less of a problem with the boundaries between layers 4 and 5 in the lower crust and in the mantle.

Therefore, partial multiples related to boundaries within the crystalline crust lead to serious noise for tracing deeper reflections. However, such noise occurs simultaneously with single reflections only in areas where the angle of incidence is relatively large, near the critical angle. Calculated travel-time curves for multiply reflected waves for this zone show that partial multiples at these distances have a lower apparent velocity than single reflections from deeper boundaries [70]. As a consequence, the controlled directional reception recording method should be very useful in separating useful waves from multiples.

In addition to multiples with a travel time equal to that for a single reflection in a layered medium, many ghost events formed as shown in Fig. 32a for a partial multiple may arise. Such waves also hamper the identification of deeper reflections.

The intensity of ghosts is always lower than that of a single reflection from the same boundary. However, with horizontal stratification, the case may arise for which partial reflections on entry and exit may add. This case will lead to a ghost event being stronger than the primary event only in the case $K_{pr} > 0.5$, with a single intermediate boundary having the reflection coefficient K_{pr}. Such a condition may occur when the angle of incidence is close to critical. For steeper angles and continental conditions $K_{pr} < 0.5$.

In the case of an oceanic crust, all of the partial multiples are much stronger than primary events because of the large reflection coefficients involved. This is also true of multiple reflected-refracted events. In deep seismic soundings at sea, the fifth and seventh multiples of such events may readily be seen ([163], Chapter 3).

The evaluation which has been presented here is highly simplified and only approximate, but it does indicate that deep reflections close to the shot point must be much weaker than in the vicinity of the critical angle, where the principal noise for distances up to 50 km and times of 10 to 20 sec consists of low-velocity waves which may be effectively suppressed with geophone patterns, provided the noise is no more than 3-4 times as strong as the deep primary events.

Directly over the shot point in basement outcrop areas, the principal noise interfering with detection of deep events consists of nonstationary noise associated with the shot. This noise is several times more intense than the natural microseismic level.

Above we have considered one form of stationary high-velocity noise interfering with the identification of deep reflections. It is clear that in a complicated real medium, the situation is even more complex.

Problems with interfering waves are more complicated with the deep seismic sounding method than with conventional seismic exploration because of the irregular layering of the crust and its blocky structure. I suggest, therefore, that in deep seismic sounding, the use of deep reflections with angles of incidence near normal would hardly be a solution for the direct separation of useful waves from noise. When reflections do not correlate well, it is obviously necessary to use all of the events contained in a wave field simultaneously, and not just the events with vertical incidence. It will be necessary to develop equipment which would provide more reliable seismograms and enhance the deep high-velocity events. This will require a serious study of the wave forms that develop in the earth.

All of this, finally, does not exclude the possibility of frequency filtering and the use of geophone patterns in deep seismic sounding to provide the optimum conditions for identifying and tracing individual deep reflections in areas where they are well developed and may be correlated over reasonable distances.

Geophone patterns have been relatively little used in deep seismic sounding because of the cumbersome procedures involved. The introduction of magnetic recording has significantly broadened the feasibility in this regard, allowing the use of relatively small patterns in the field (3-5 geophones) with subsequent generation of more complicated patterns on replay.

§ 4. Resolution in Deep Seismic Sounding

In evaluating the resolution of the deep seismic sounding method, we will take the crust and upper mantle to be nearly horizontally layered as a first approximation. By so doing, we can make use of results developed for seismic exploration by I. S. Berzon [22-24] for the use of reflected and refracted waves in studying boundaries in a horizontally stratified sequence.

We will determine the basic parameters controlling resolution in seismic methods for the particular conditions which apply in deep seismic sounding. These parameters include the minimum vertical distance between two boundaries for which the waves from each of the boundaries may be identified separately along some interval on a traverse L.

With deep seismic sounding data, individual events are usually not traced along any significant length of a profile, but rather, only wave groups are traced in general. In order to simplify calculations, we will assume that each wave group of duration Δt is an individual event.

A. Resolution in Deep Seismic Sounding with Recording

of Refracted Waves

We will determine the conditions under which two separate refracted waves can be traced.

For a horizontally stratified medium, the beginning of the interference zone for two waves corresponding to boundaries located at depths h and h + Δh and defining media with velocities V_2 and V_3 is determined by the distance to the intersection point X_{int} of their travel-time curves [23].

$$X_{\text{int}} = \frac{2h_1}{V_1}(\cos i_{12} - \cos i_{13}) + \frac{2\Delta h}{V_1}\cos i_{23} - \frac{2h_1}{V_1 \cos i_{12}}.$$

In dealing with a multiple layered medium for the overburden above the first boundary, situated at the depth h, we may treat it as uniform with an average velocity V_1.

For well-defined records of the principal wave groups and straight-forward velocity characteristics, the distance from the shot point to the intersection point on the travel-time curves for the M and K wave groups is determined primarily by the depth to the M-discontinuity. This dependence is used in adjusting observed data during field operations.

The extent of the interference zone ΔX_I is determined by the expression [23]:

$$\Delta X_I = \frac{a_1 + a_2}{\frac{1}{V_2} - \frac{1}{V_3}},$$

where a_1 and a_2 are the pulse lengths for each of the interfering waves.

If we assume that the length of each wave group changes but little with distance from the source and the length is essentially the same for both wave groups, we have

$$\Delta X_I = \frac{2a}{\frac{1}{V_2} - \frac{1}{V_3}}. \tag{3.1}$$

The extent of the interference zone ΔX_I is indicated in Table 9 for the conditions applicable to deep seismic sounding. Values for ΔX_I were calculated for various combinations of wave groups: sedimentary and crustal, crustal and mantle. Respectively larger durations a_1 and a_2 were used for these cases. In computing the interference zone for individual waves within wave groups, average values for the durations a_1 and a_2 observed in various areas where field work has been done at frequencies of 5-10 Hz were used.

Equation (3.1) may be used in computing the duration of the interference zone for two reflected waves at angles greater than critical, where their travel-time curves are similar in character to those for refracted waves.

If we consider interference between two reflected waves from boundaries at depths h_1 and h_2, over which the velocities are, respectively, V_1 and V_2, it is possible to assume as a reasonable approximation that the apparent velocities of the reflected waves are V_1 and V_2, respectively. If $V_3 > V_2$, with V_3 being the velocity below the lower boundary, the interference zone between the two reflected waves for greater than critical angles of incidence will always be smaller than the interference zone for two waves with velocities V_3 and V_2 refracted from the same boundaries. It is apparent from Table 9 that because of the small velocity differentials in the crust, the interference zones are of considerable extent and identification of individual waves is difficult.

However, the dimensions of actually observed interference zones are fifty percent smaller than the computed values as a result of pulse group shortening caused by the attenuation of the leading waves with distance from the shot point and the difficulty in seeing these events above the background noise level (see Chapter II). Moreover, with significant differences in the intensities of the interfering waves which are encountered in many areas, conditions develop which are favorable to the continuous tracing of the dominant wave.

Because of the nature of the velocity profile in the crust and mantle, refracted waves or reflected waves at greater than critical angles from deep boundaries are relatively more intense at a given distance from the shot point. This phenomenon is also favorable for identify-

TABLE 9. Extent of the Interference Zone for Crustal (K)
and Mantle (M) Waves

Interfering waves	Velocity, km/sec		Length of wave group, sec		ΔX_I, km	
	V_2	V_3	a_1	a_2	calc.	observed
Sedimentary and crustal waves (S and K)	4	6	0.5	2.0	30	20
Interference between waves in the K group	5.5	6.2	0.5	0.5	50	—
	6.0	6.6	0.5	0.5	66	20—30
	6.6	7.2	0.5	0.5	80	—
Interference between K and M waves	7.2	7.6	0.5	0.5	77	—
	6.6	8.2	1.5	1.5	100	50
Interference between waves in the M group	8.2	8.6	0.5	0.5	114	—
	8.6	9.0	0.5	0.5	194	—

ing deep waves in the later parts of a record and tracing them across interference zones with weaker, shallower waves.

An expression for determining the minimum thickness of a layer Δh for which two separate waves of length a sec can be traced on either side of the interference zone is given in [23].

Only the first part is of interest to us, because after the interference zone, deep events with high velocity are traced as first arrivals, and interference is much less of a problem. Low-velocity waves in most cases are strongly attenuated after travel to this region, and their recognition is difficult.

We will establish the conditions for recognition of waves in a given interval L along a traverse. We will consider the case where the difference between the abscissas X_{crit} of the initial (critical points) for two interfering waves is small; i.e., $X_{crit\,1} - X_{crit\,2} < $ L and the initial point for the second wave is found at a distance less than the distance X_{int} to the intersection of the travel-time curves. This condition usually applies in the case of deep seismic sounding:

$$L = X_{int} - \Delta X_I - X_{crit}.$$

We can obtain an expression from (3.2) for computing the thickness Δh for which refracted waves will be recorded separately:

$$\Delta h = \frac{V_3 - V_2}{V_3 \cos i_{23}}\left(h \tan i_{12} + \frac{L}{2}\right) + \frac{aV_2}{2\cos i_{23}} - \frac{hV_2}{V_1 \cos i_{23}}(\cos i_{13} - \cos i_{12}),$$

$$i_{12} = \text{arc sin}\frac{V_1}{V_2}, \qquad i_{23} = \text{arc sin}\frac{V_2}{V_3},$$

where a is the length of a wave group.

We will investigate this relationship for the conditions of deep seismic sounding for various h, V_1, V_2, V_3 and, for wave durations a = 0.25, 0.50, and 1.0 sec.

Computations were done for four typical crustal sections:

	1	2	3	4
V_1	6.0	6.0	6.0	6.5
V_2	6.5	6.5	7.0	8.0
V_3	8.0	7.0	8.0	8.5

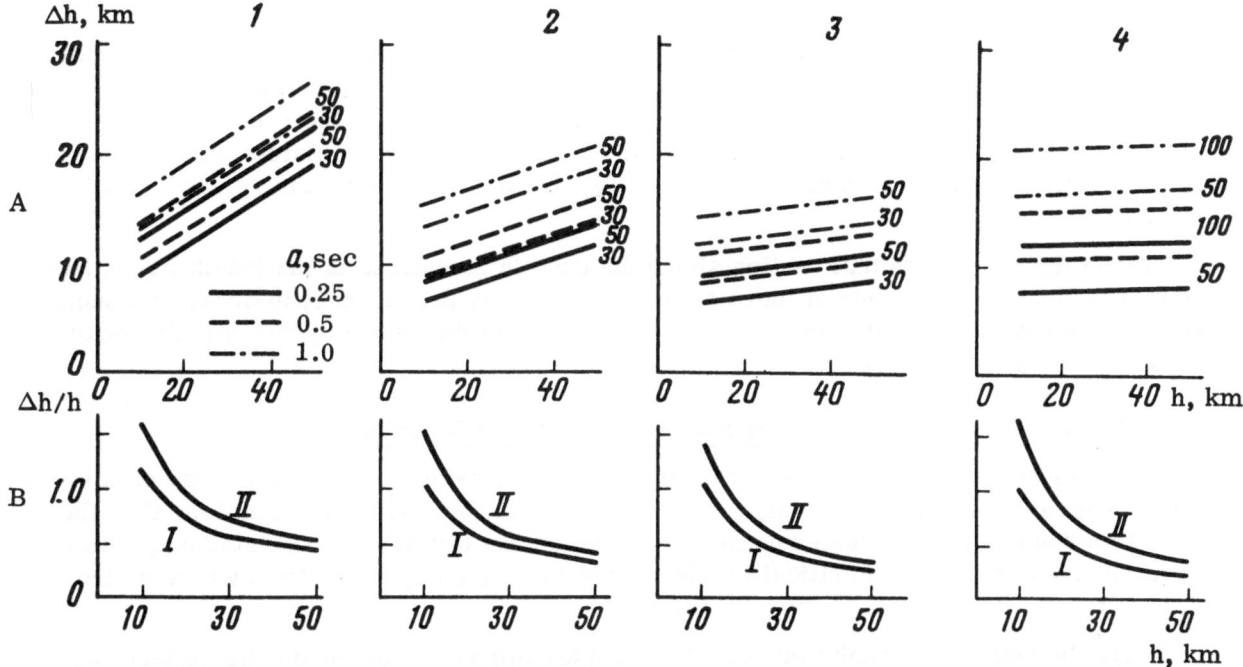

Fig. 3.3. Resolution with the use of refracted waves in deep seismic sounding: A) Δh is the absolute resolution, with the curve parameter being L, in km; B) relative resolution; I) L = 50 km and a = 0.5 sec; II) L = 50 km and a = 1 sec; Δh is the thickness of a layer for which two individual refracted waves can be recorded in an interval L; h is the depth to boundary. For the properties of each section, see text.

Curves for the relationship between Δh and the depth h are shown in Fig. 3.3. They have been calculated for various values of a and for L = 30 and 50 km. Curves for the relative resolution Δh/h are given in the lower part of the illustration for L = 50 km and a = 0.5 sec (curve I) and a = 1.0 sec (curve II).

The values for Δh increase with depth for all of the sections in Fig. 3.3. The rate of increase differs for the different sections, but for a given section, the curves for various values of a and h are all parallel.

The smallest values for Δh, which are nearly independent of h, are found with sections 3 and 4, characterized by boundaries in the lower crust and upper mantle. With a = 0.5 sec and L = 50 km, Δh ranges from 10 to 12 km as the depth to the boundary changes from 20 to 50 km.

For intermediate boundaries in the crust (sections 1 and 2), with the same values for a and L, Δh varies from 14 to 22 km.

Thus, the absolute resolution is poorer with greater depth to a boundary, but is better for the intermediate boundaries in the crust than for boundaries in the lower crust and upper mantle.

As may be seen from the lower series of curves, the relative resolution Δh/h behaves consistently. It improves for deeper boundaries for all of the sections. The largest rate of decrease in Δh/h (which corresponds to the most rapid rate of improvement in resolution) is observed for the range 10-30 km; all the curves flatten out at greater depths (30-50 km).

The best relative resolution is noted for sections 3 and 4. It amounts to 0.2-0.3 at depths of h = 30-50 km. For sections 1 and 2, at the same depths Δh/h is 0.4-0.6.

In evaluating resolution, it is important to consider the effect of the following factor: an increase in group length from 0.25 sec to 1.0 sec, which amounts to a fourfold reduction in frequency if the same number of oscillations are included in a group, results in an increase of roughly 50% for each of the four sections (1.3 to 1.8). A decrease in the recording interval L from 50 to 30 km, which requires about twice the density of recording points, causes a decrease in Δh, or an improvement in resolution by 30-40%, but the improvement in relative resolution is no more than 10%.

These figures stress that caution should be used in investigating the feasibility of improving resolution with the use of more complicated survey procedures, in order that such modifications not lead to some other requirement, such as the necessity for simultaneous tracing of several boundaries.

B. Resolution in Recording of Reflections in Deep Seismic Sounding

In distinction to routine seismic exploration (where reflections are recorded almost entirely at angles steeper than critical, usually at distances from the shot point less than the depth h to a boundary) with deep seismic sounding, as has already been pointed out, better reflections are obtained with critical incidence and at angles beyond critical where reflections are close to refractions, that is, usually for x > 2h.

Thus, in evaluating resolution, we will consider two areas for recording reflections: less-than-critical reflections at points near the shot point, where travel-time curves for deep reflections are nearly parallel to one another and conditions for their interference are determined only by the length of a particular oscillation in a wave group, and, as shown in [22, 23], are not dependent on the depth to a boundary; and more-than-critical reflections, for which travel-time curves are similar in form to those for refracted waves. In this region, the conditions for interference of two reflected waves are essentially the same as the conditions for interference between two refracted waves.

1. Area of "Vertical" Reflections. Reflections with angles of incidence close to vertical, with distance from the shot point not exceeding h/2, will be recorded separately under the condition

$$\Delta h \geqslant \frac{a\overline{V}}{2},$$

where $\overline{V}$ is the average velocity between the depths h and h + Δh, and a is the length of a group.

For boundaries in the crust

a, sec	$\overline{V}$, km/sec	Δh, km
0.25	5.5	0.7
0.5	5.5	1.4
1.0	6.0	3.0
1.0	6.4	3.2

For boundaries in the mantle

0.5	7.2	1.8
1.0	7.2	3.6
0.5	8.0	2.0

Thus, we see that the resolution for vertical reflections in the seismic method remains quite good, even for long groups of oscillations, being 3-4 km.

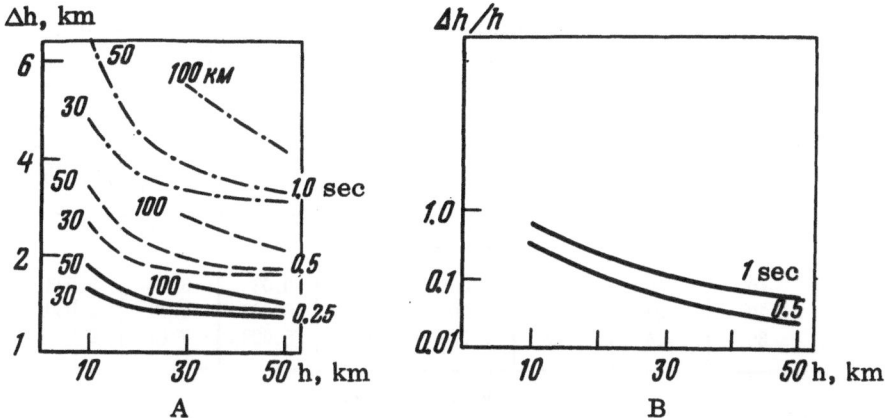

Fig. 3.4. Resolution with reflected waves in deep seismic sounding:
A) Δh is the absolute resolution (the figures to the left are the
lengths of the interval L in km, and the figures to the right are the
lengths a of a wave group); B) relative resolution for L = 50 km;
the lengths a are indicated on the curves; the velocity in the crust
is 6 km/sec.

These figures illustrate quite well the limits of resolution which might be obtainable in
deep seismic sounding with the recording of clear, nearly vertical reflections. These figures
may also be used to judge the limiting thinness of a layer between two reflecting boundaries
which may be resolved in deep seismic sounding.

2. Conditions for Distinct Recording of Two Reflections at Different Distances from the Shot Point. We will consider the conditions for distinguishing reflected waves over an interval L on a profile.

We will determine the minimum value of Δh for which two reflected waves can be recorded separately, beginning at the shot point and extending to a distance x = L km. The value
for Δh in this case, as in the case of refracted waves, will depend on the properties of the
medium — the average velocity and the depth to the layer.

We will assume that the velocity $\overline{V}$ is constant in the medium, and that the depth to the
boundary h varies from 10 to 50 km. We will compute Δh and Δh/h for L = 30-50 and 100 km
for a medium with $\overline{V}$ = 6 km/sec (Fig. 3.4). The duration a of the wave varies as follows:
a = 0.25, 0.5, and 1.0 sec. The first of these values for a corresponds to individual waves
within a group, while the second and third values correspond either to groups of waves or to
single waves at seismological frequencies (2.5 Hz).

The curves in Fig. 3.4 were computed using the expression:

$$\Delta h = \frac{1}{2}\sqrt{\left(aV_1 + \sqrt{x^2 + 4h^2}\right)^2 - x^2} - h \ . \tag{3.3}$$

In calculations with Eq. (3.3), values for x were taken equal to L. The values for Δh
decrease nonlinearly with the depth h; for L = 30-50 km, the decrease is less rapid than at
lesser depths (10-30 km), and becomes even less rapid at greater depths. For L = 100 km
and depths of 30-50 km, the relationship between Δh and h is nearly linear. The value for Δh
with L = 50 km and a = 0.5 sec is 3.5 km for h = 10 km and only 1.5 km for h = 50 km.

The curves are nearly parallel for various values of L and a, and so we may say that
for the range of depths being considered, there is a proportional improvement in resolution
with increasing frequency.

TABLE 10. Resolution of Reflected Waves in the Vicinity
of the Initial Point (Critical Angle)

h_1, km	$L = X_{crit}$, km	a, sec					
		0.25		0.5		1.0	
		Δh, km	$\frac{V'_2 - V_1}{V_1}$*	Δh, km	$\frac{V'_2 - V_1}{V_1}$	Δh, km	$\frac{V'_2 - V_1}{V_1}$
$V_2 = 6.5$ km/sec							
10	48	1.81	0.008	3.42	0.02	6.30	0.028
20	96	1.86	0.005	3.63	0.003	6.70	0.020
30	114	1.86	0.005	3.77	0.008	7.18	0.012
$V_2 = 7.0$ km/sec							
20	69.4	1.45	0.008	2.85	0.015	5.48	0.030
30	104.4	1.46	0.005	2.90	0.013	5.64	0.025
40	138.8	1.50	0.005	2.96	0.007	5.75	0.017
$V_2 = 8.10$ km/sec							
30	67.8	1.10	0.010	2.15	0.017	4.34	0.033
40	90.4	1.10	0.005	2.20	0.012	4.40	0.025
50	113	1.20	0.003	2.30	0.010	4.50	0.022

*$V_1 = 6.0$ km/sec. V'_2 is the average velocity to the lower boundary considering the layer with thickness Δh.

The relative resolution for reflected waves increases with depth relatively more rapidly than does Δh. For L = 50 and a = 0.5 sec it amounts to 34% for h = 10 km, while it is only 3.5% for h = 50 km. It decreases by a factor of nearly two for a change in h from 30 to 50 km.

Thus, even at relatively large distances from the shot point (up to 50 km), the resolution for reflections at less than critical angles remains high. We will now consider how it changes as we approach the region of the initial point, where the travel-time curve for reflected waves is similar in form to the travel-time curve for refracted waves.

3. Separate Recording of Reflected Waves in the Region of the Initial Point for Head Waves. This case is of interest for determining the minimum value of Δh for which two reflected waves can be recorded separately in the most favorable, with respect to signal strength, region.

Calculations were done using the approximate expression (3.3); in so doing, the same assumptions about average velocity as in the preceding case were used. The results are given in Table 10, which lists values for the spread between V_1 for the first boundary at a depth h, and V_2 for the second boundary located at a depth h + Δh.

In (3.3), it was assumed in the computations that x = L − X_{crit} where X_{crit} is the abscissa of the initial point.

Computations of Δh for three different sections with different values for the velocity V_2 are given in Table 10. These cases correspond to various boundaries in the crust. With a change in depth to the boundary from 10 to 50 km, the distance L to the initial point changes from 50 to 140 km. For each of the sections, computations were done for a variety of depths corresponding to values observed for a continental crust.

For a = 0.5 sec, the values for Δh vary little with increasing h, that is, with greater depth of burial of the interface. For the first section considered, Δh varies over the range 3.4-3.8 km; for the second, 2.85 to 2.95 km; and for the third, from 2.15 to 2.30 km. Thus, in the region of the critical angle, the absolute resolution is almost independent of V_2 and h. The value for Δh is essentially equal to aV. The relative resolution will increase with depth in inverse proportion to the ratio ΔV/h.

C. Comparison of Δh and Δh/h for Reflected and Refracted Waves

We will now compare the resolution obtainable with reflected waves and with refracted waves in deep seismic sounding.

The ranges of variation of Δh and Δh/h for reflected (refl) and refracted (refr) waves are indicated on Fig. 3.5 for L = 50 km and a = 0.5-1.0 sec; the choice of a = 1.0 sec corresponds to the frequency used in seismological profiling and in low-detail deep seismic sounding data. A comparison of the ranges for reflected and refracted waves on the curves for Δh and Δh/h indicates that:

1. For the conditions applicable to deep seismic sounding, resolution with the use of reflected waves is 3 to 10 times better than obtained with the use of refracted waves; with the use of reflected waves at less than critical angles and at critical angles, layers as thin as 2-3 km may be recognized at great depth (30-50 km), or layers as thin as 3-5 km may be recognized at shallower depths (10-20 km); with the use of refracted waves at L = 50 km, the least thickness of a layer which can be detected at depths of 30-50 km is 10 km;

2. There are significant differences in the resolution of deep seismic sounding for the study of boundaries at various depths; the most favorable conditions are found for depths of 30-50 km; at these depths the relative resolution with the use of refracted waves at L = 50 km and a = 0.5 sec is 0.2-0.4, and with reflected waves 0.03-0.07; at lesser depths (10-20 km) the resolutions are lowered, respectively, to 0.5-1.0 and 0.3-0.6; that is, at the shallower depths, refracted and reflected waves are comparable; with transition to higher frequency and lesser L, the relation is even clearer.

The evaluation of the resolution obtainable in deep seismic sounding for the simplest case of the recording of two refracted or two reflected waves is relatively formal and free from consideration of seismic-wave dynamic characteristics. An extension of the analysis to a crust with many layers involves major complications.

In the first place, there are not just two boundaries in the crust and mantle, but a number of boundaries at which reflections and refractions develop, with mutual interference between all of these waves. This further complicates the problem of tracing each of the waves. However, there are a number of factors which favor the identification and tracing of waves through interference zones. Among these are the attenuation of waves with distance from the source, which causes waves to disappear, within certain dynamic ranges, and the existence of dominant waves, which are characteristic of specific velocity profiles.

With the use of continuous profiling, these last two factors are widely used in deep seismic sounding for tracing waves through interference zones. This permits a lengthening of the interval over which a wave is recorded, and increases the resolution so that as the methods are being used at present, resolution obtained with refracted waves is better than that calculated above. Deep seismic sounding has been used in many instances to recognize layers with a thickness of less than 5 km. Thus, a high degree of resolution has been obtained with continuous profiling using a high station density, allowing recognition of individual events with a duration a = 0.25 sec, tracing of these events over moderately short intervals of L < 50 km and use of travel-time curves for reflected and refracted waves in interpretation.

The analysis of resolution presented here has indicated that despite the fact that resolution is determined primarily by the frequencies used, obtaining this resolution is possible only

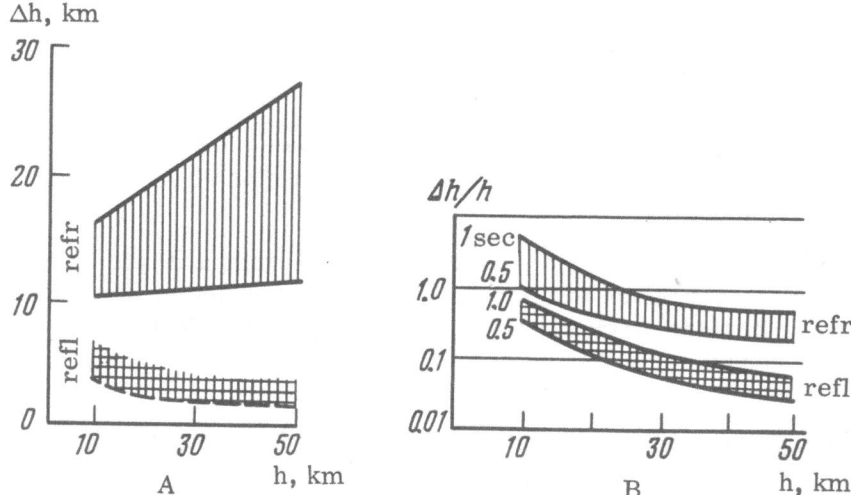

Fig. 3.5. Comparison of resolution obtainable with reflected and refracted waves in deep seismic sounding: A) Δh is the absolute resolution for $L = 50$ km and $a = 0.5$-1.0; $L = 30$ km for reflected waves; B) the same, in terms of relative resolution.

with the use of a survey system of travel-time curves with sufficient accuracy for constructing a section.

In practically all branches of seismology, including both deep seismic sounding and conventional seismic exploration, more events can be recognized on individual records than can be correlated between records. In deep seismic sounding, this discrepancy is particularly notable with the use of piecewise continuous profiling. Use of frequencies of 8-12 Hz allows recognition of individual events with a duration of 0.25 sec or wave groups with a duration of 0.5 sec along the individual segments of a piecewise continuous profile. However, in correlating events between different segments, we must resort to group correlation; that is, waves must be grouped into intervals which are rarely shorter than 1-2 sec reducing the resolution to that which one would expect for a frequency of 2-5 Hz.

In addition to continuous profiling, in which, in principle at least, it is possible to obtain the ideal resolution for the frequencies used (8-12 Hz), reasonably good agreement between frequencies and the distances between observation points is obtained with point soundings at sea. With records made using the frequency range 4-6 Hz, it is necessary to use correlation of wave groups with a duration of about 0.5 sec when the separation between shot points is 3-5 km.

The density of observation points is somewhat inadequate for these frequencies, so that commonly in marine surveys, data do not permit the construction of as reliable sections as might be possible.

Further improvement in the resolution of deep seismic sounding in terms of the extent to which the crust and mantle can be subdivided may have to be based on the simultaneous recording of reflected and refracted waves and the extension of interpretation to later waves in wave groups, using wave-form parameters not now recognized, primarily the spectrum of a wave, for broadening the applicability of the method. Development in these directions still requires serious study of the physical basis for the method and study of the nature of reflection and refraction at boundaries in the crust and mantle.

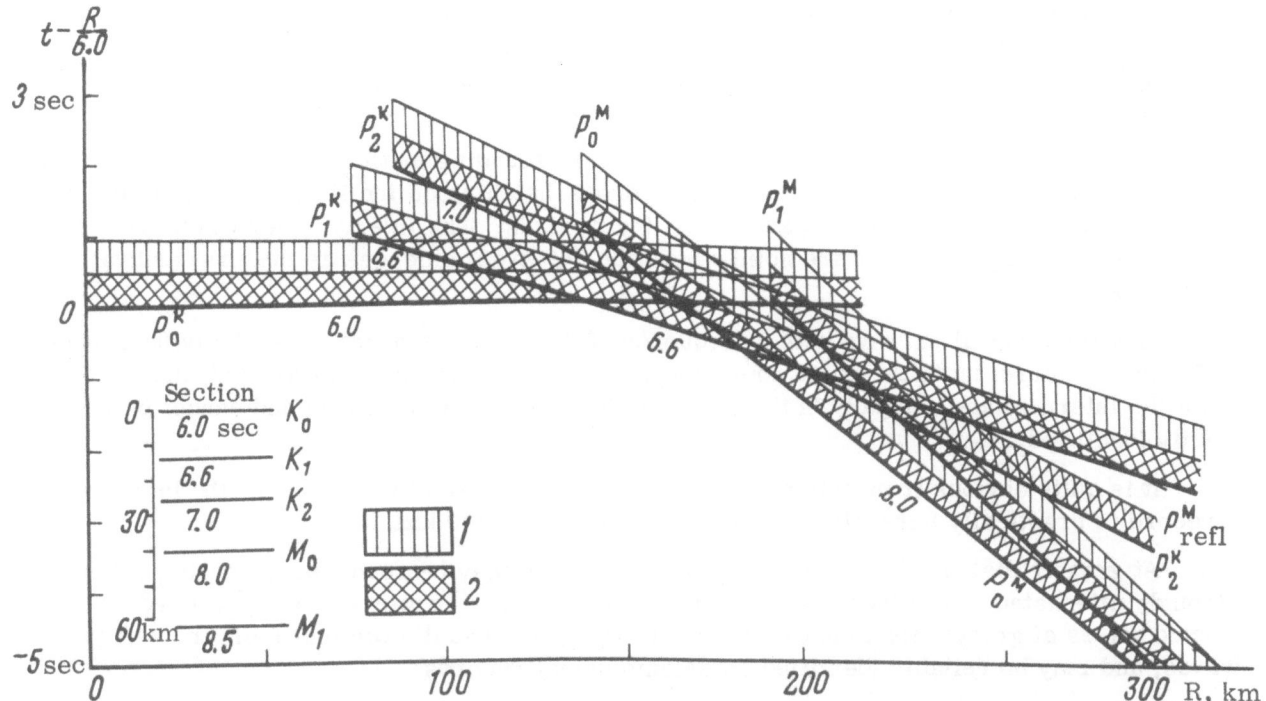

Fig. 3.6. Interference regions for P^K and P^M waves for various durations of wave groups: 1) for seismological profiling; 2) for deep seismic sounding. The figures on the section indicate velocity in km/sec.

§ 5. Applicability of Deep Seismic Sounding and Seismological Profiling to Subdivision of the Earth's Crust

The computations of resolution in deep seismic sounding were carried out for frequencies ranging from 2.5 to 10 Hz. Wave durations of 1.0 sec corresponds to the conditions encountered in seismological profiling. Therefore, any of the results applicable to recording reflections and refractions in deep seismic sounding is pertinent to the conditions in seismological profiling, with the point survey method being considered specifically. Using this approach, we will compare the applicability of deep seismic sounding and seismological profiling from the point of view of subdividing the crust into individual layers.

The practicality of identifying layers, as we have seen, is determined by the practicality of recording separate events corresponding to the top and bottom boundaries of a layer. The ability to separate events, in turn, depends on the nature of the method — the frequencies used and the survey procedures.

A comparison of the curves in Figs. 3.4 and 3.5 indicates that the resolution of the seismological profiling method (considering only the recording of longitudinal waves along radial profiles) is lower by a factor of two or three than that for the deep seismic sounding method; that is, essentially, the method cannot be used to identify layers within the crust that have a thickness of less than 10-15 km. In principle, thick layers may be recognized with this method, but the reliability of tracing such layers along a survey profile, with distances of 10 km or more between recording stations, is complicated by interference with other waves.

For clarification, we can examine the travel-time curves for the primary K and M waves (see Fig. 3.6). In the model postulated for the crust, there are two intermediate boundaries defining layers with velocities of 6.6 and 7.0 km/sec. The velocity in the upper part of the crust is 6 km/sec, while below the crust it is 8 km/sec.

With observations as made in the deep seismic sounding method with a duration for wave groups of 0.25-0.50 sec (shown by the heavier cross-hatching on the illustration), almost all of the events may be identified separately on the records. The interval over which the P_2^K event may be recognized up to the point where the P_0^M event emerges as the first arrival is about 30 km.

In seismological surveys with a frequency of 2.5 Hz and a duration for individual events of 1.0 sec or more, up to the point where the P_0^M wave emerges as first arrival, the P_2^K event from the intermediate boundary in the crust is obscured by interference and cannot be reliably identified.

It is quite obvious from this example that at seismological frequencies, the later part of a record is practically unresolvable over distances of 50-200 km.

At greater distances, where the P_1^K and P_2^K waves may be recorded separately, they are strongly attenuated, while there is interference between the mantle waves P_0^M and P_1^M. Reflected waves at greater than the critical angle P_{0refl}^M are usually distinct from the refracted waves, and may be reliably identified on seismological records.

We will return our attention to the basis of the complexity of the interference zone for mantle waves. These form a compact group of partially reflected-partially refracted waves, which have dynamic characteristics that are difficult to separate over long correlation intervals.

In cases where the M events may be recognized on the records, as for example, may be the case when earlier events are attenuated, these may be used to determine the structure of the upper mantle.

Inasmuch as for boundaries at depths of more than 30-40 km the resolution depends only slightly on changes in the parameters describing the section or further increase in the depth h, using Figs. 3.3 and 3.4 along with Table 10, it is possible to determine the thickness of a layer in the mantle which may be studied using travel-time curves for reflected and refracted body waves recorded in deep seismic sounding or seismological profiling surveys:

Method	Δh, km
Deep seismic sounding, reflection	2-5
refraction	8-15
Seismological profiling, reflection	4-8
refraction	15-20

Layers with essentially these thicknesses have been identified in the upper mantle from travel-time curves for refracted waves and reflected waves at less than critical angles in operations with the deep seismic sounding method (Ryaboi [153]) and with seismological profiling (Lukk and Nersesov [126]) along long profiles.

In conclusion, we will present some data on the resolution of seismic methods for studying the crust in comparison with the recorded wavelengths.[†]

[†] A similar evaluation has been made by Nikolaev in "Seismic properties of soils," Izd. Nauka (1966).

TABLE 11. Comparison of Resolution of Seismic Methods Used
in Crustal Studies

	Deep seismic sounding	Seismological profiling	Surface-wave dispersion
Frequency, Hz	5-10	1-3	0.01-0.03
Wavelength λ in the crust, km	1.2-0.6	6-2	350-100
Velocity	V_p = 6 km/sec		V_R = 3.5 km/sec
Thickness of the crust in wavelengths (H = 10-40 km)	20-100	2-30	0.03-0.01
Minimum thickness of a layer, expressed in wavelengths	10-15 (refr) 3-5 (refl)	3-5	0.01

Table 11 shows the relationship between wavelengths and crustal thickness for the deep seismic sounding method, the seismological profiling method and the surface-wave dispersion method. It is interesting to note that in seismological surveys and deep seismic sounding surveys in which body waves are used, layer thicknesses which may be recognized range from a few wavelengths to tens of wavelengths in thickness, while with the surface-wave dispersion method, layers are identified with a thickness of only a hundredth of a wavelength. This phenomenon is explained by the fact that in the surface-wave dispersion method supplementary discrimination is used in the form of a relationship between velocity and frequency, which enhances the resolution of this method.

It is also of interest to compare the degree of horizontal inhomogeneity which may be recognized with the various methods. However, this may be done only rather crudely, as the theoretical basis for these calculations is poorly developed.

On the basis of deep seismic sounding results in various areas, we may conclude that individual crustal blocks characterized by specific velocity profiles down to the M-discontinuity may be recognized reliably if the dimensions of such blocks are comparable with the length of a travel-time curve segment, i.e., no less than several tens of kilometers (up to 100 km and more).

With a dense station network in seismological profiling, crustal blocks characterized by different depths to the M-discontinuity can be recognized. The dimensions of such blocks are of the order of 100-200 km and more [110].

In the study of crustal structure using body waves, crustal blocks with dimensions much larger than a few hundred kilometers, which corresponds to a hundred wavelengths, may be recognized.

With the surface-wave method, horizontal inhomogeneities with dimensions of a few hundred kilometers, corresponding to a wavelength, are studied.

Thus, even from the point of view of the resolution for horizontal inhomogeneities, the same relative positions between the body-wave and surface-wave method holds. The absolute resolution obtainable with body-wave methods is always much better than that obtainable with surface-wave methods because of the shorter wavelengths (see Chapter I).

The physical nature of these differences underlines the importance of studying new wave parameters, properties of the medium and wave spectra which may result in an improvement in the resolution of methods for the investigation of the crust and mantle based on the recording of body waves.

CHAPTER IV

WAVE PROPAGATION AND MODELS FOR THE MEDIUM

With shooting at a single point and observation of the resultant seismic vibrations at many points, wave propagation may be described functionally as $A = f(t, R)$, where A is the "intensity" of the vibration, t is time, and R represents the spatial coordinates x, y, and z. Here, A is a value representing the amplitude of any wave (or part of it) under equivalent conditions. It is a scalar, vector, or tensor quantity: displacement, particle velocity, acceleration, pressure, stress, strain, and so on. In the simplest case, it may be assumed that A and R are scalar quantities, with A being the deflection on a record and R the distance from the shot point to the observation point.

The primary elements in a seismic wave field are the regular waves, for which in the simplest case and with some approximation, $A = f(t - R/a^*)$, where a^* is the speed of propagation of a particular wave in the R direction.

As we have seen in Chapter II, the regular waves generated by an explosion are characterized by narrow or resonant spectra, so that they exhibit a quasi-sinusoidal form of limited duration on a record. A wave usually consists of 2-2.5 oscillations (Fig. 4.1). In interpretation, regular waves are formed by longer groups of quasi-sinusoidal waves with an elongate and complicated form.

The regular waves are accompanied by irregular noise, weak vibrations, which form the seismic noise. Considering the regular waves and the noise between these waves for the interval over which observations are made, there is a consistent variation of time t with distance. However, the rate of change varies and depends on the structure of the crust.

The analysis of a wave field is fundamental to the interpretation of deep seismic sounding data, which has been done essentially as follows in recent years:

1. waves are identified and correlated to form travel-time curves;

2. the velocity structure of the medium is determined and a section of the crust and upper mantle is constructed, assuming plane layering;

3. wave speeds and wave forms, computed from this model, are compared with the actually observed events; discrepancies between the computed and observed data are analyzed, and a new model, which includes velocity inhomogeneities in the crustal and mantle layers, is developed;

4. the results so obtained are reconciled with geological and other geophysical data.

The primary step in this scheme is the correlation of events between records. The idea of correlation of wave groups and groups of parallel travel-time segments as well as of individual events has been developed in deep seismic sounding (Gamburtsev [50]; Veitsman [36]).

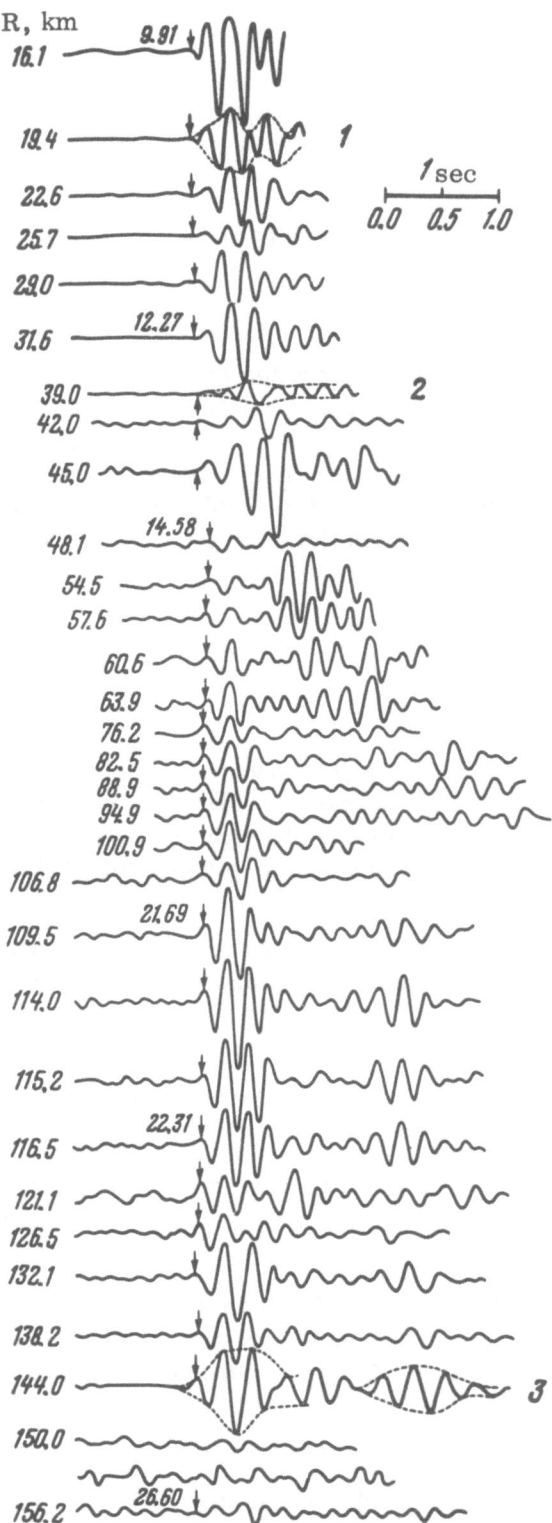

Fig. 4.1. Examples of records made at a single recording site in the Pacific Ocean, 1964 (Institute of Physics of the Earth, Zverev and others). The numbers to the left are distances in km; the arrows indicate the time of the first arrival; the dashed curves show elongate wave groups.

Further operations with groups of parallel travel-time segments are done under the supposition that each group has properties which are identical for waves of a given type, i.e., they are longitudinal or dilatational, simple or complex (multiples), reflections or refractions, and so on.

The basis for, or in the last viewpoint, the justification for such an assumption required careful study of the subject of wave propagation. This became practical only with the wide use of continuous profiling and a high station density in deep seismic sounding.

Now, we will consider the large volume of data on the nature of wave propagation which is available for the various parts of the USSR.

Analysis of these data and comparison with the results from less detailed deep seismic sounding surveys, as well as with the results from seismological studies, will permit us to consider the question of the physical basis for correlation of travel-time segments from a new viewpoint, to evaluate the error properties, and to specify what should comprise a correlatable wave group for survey systems with various degrees of detail.

§ 1. Properties of Records, Identifiable and Correlatable Waves for Various Forms of Profiling

Depending upon the survey procedure used in deep seismic sounding, various types of records — seismograms — are obtained. For continuous profiling, segmental profiling, and even for point soundings on land, usually multi-channel recording is used, providing records similar to those obtained in standard exploration, on which regular waves may be identified by correlation between adjacent traces. Generally this correlation defines a stepout.

In observations as they are presently made at sea, only a single recording channel is used, and the records are analogous to seismograms obtained at a seismic observatory. The identification of specific regular waves on such records (the records may consist of several traces, when several filter settings are used) can be based only on wave form.

A comparison of single-channel and multi-channel records for observations on land and at sea [3, 47, 103, 163] provides a basis for identifying specific quasi-sinusoidal oscillations as wave events on the basis of amplitudes (Fig. 4.1).

Depending upon the form of profiling, different principles are employed for correlating (or tracing) events on a group of records.

1. With continuous profiling and observations along a specific part of a profile, correlation of individual phases and waves is widely used, and the principles for this have been described thoroughly in several texts on seismic exploration [48, 49, 59, 135]. However, the nature of the records obtained in deep seismic sounding does not allow this step to be used indiscriminately.

On multi-channel deep seismic sounding records at a frequency of 10-15 Hz, a large number of distinct regular events (up to several tens) may be recognized. As has been found from an analysis of records obtained in various areas, these events may be correlated without trouble only over relatively short distances along a profile (no more than a few tens of kilometers). Then, correlation of a particular event is lost, and another event with a travel-time and apparent velocity close to that of the preceding event can be traced. Sometimes a shift in a wave, related to a change in velocity, is observed, which is analogous to the shift seen with a horizontally layered medium. Frequently these events bear a character typical of a medium with vertical boundaries (scatter of amplitudes, rapid attenuation, correlation of events only over a few adjacent traces, and so on). With shifted waves, in correlating suc-

TABLE 12. Correlation Intervals for Specific Waves in Groups, for Continuous Profiling

Wave type	Southwestern Russian Platform [131]				Bukhara–Khivin region [42]				Southwestern Turkmenia [42]				Dzhezkazgan region [134]			
	wave	V, km/sec	range of identification, km	range of correlation, km	wave	V, km/sec	range of identification, km	range of correlation, km	wave	V, km/sec	range of identification, km	range of correlation, km	wave	V, km/sec	range of identification, km	range of correlation, km
Crustal K	P_1^0 (P_0^K)	5.8—6.4	30—55	4—8—40	P_z (P_0^K)	6.0—6.2	120	5—20	$P_{\mathrm{H-T}}$ (P_0^K)	6.2—6.4	35	35	$t_{1,2}$ (P_0^K)	5.8—6.1	50	50
	P_2^0 (P_1^K)	6.6	~60	40—60	P^0 (P_1^K)	6.7	70	20—25	P_z (P_1^K)	6.5—6.7	70	10—15 to 70	P_1^0 (P_1^K)	6.0—6.2	50	50
	P_1^* (P_2^K)	—	80—120	5—15 (Rarely 25)	T (P_2^K)	—	100	4—25	P^0 (P_2^K)	6.9—7.1	70	10—15 to 70	P_2^0 (P_2^K)	6.0—6.3	85	85
	P_2^* (P_3^K)	—	20—60	5—15 (Rarely 30)					P^* (P_3^K)	7.4—7.6	100	6—15 to 35	P_3^0 (P_3^K)	6.4—6.5	80	80
													P_1^* (P_4^K)	6.7—7.2	120	до 40
													P_2^* (P_5^K)	7.3—7.4	135	5—30
Mantle M	P_1^{refl}		130—190	10—60 and more	P_{refl} (P_{refl}^M)		260	8—50	P_{refl} ($P_{0\mathrm{refl}}^M$)		230	10—30	P_{refl} ($P_{0\mathrm{refl}}^M$)		120	5—60
	P_1^{pr}	8,1	80—60	2—5	P_{pr} (P_0^M)	8.1 8.3	130	4—10	P_{pr} (P_0^M)	8.2—8.4	50	4—12				

Note: Wave indexes listed first in the table are those utilized by the authors, those listed in brackets are the indexes newly adopted by the Deep Seismic Sounding Section of the Earth-Science Division of the Academy of Sciences of the Soviet Union.

cessive travel-time segments, the shift may be either toward events later in time or toward events earlier in time. Individual waves which have similar values of apparent velocity, transit time, amplitude, and wave form may be combined to form a group. Groups formed as combinations of similar waves may usually be identified and correlated over greater distances along a continuous profile. The range of recording for the principal groups of waves may amount to several tens of kilometers, as indicated in Table 12.

Correlative wave groups and groups of correlated events have essentially the same properties along a profile as correlative single events: recorded wave forms, dominant velocity and frequency, transit time, and relative intensity in comparison with other groups of waves. These characteristics are used to identify a group of waves recorded on separate segments of a segmentally continuous profile. Detailed surveys are of considerable value for combining waves into groups. It is obvious that the more detailed the survey is, the greater will be the number of wave groups that can be identified and traced.

In continuous profiling over the distance range from 50 to 300 km, usually 5-10 or even more wave groups may be recognized [42, 69, 131, 153]. In segmental profiling or point soundings, the recognition of waves as groups is more difficult, and, as is well known, only 3-4 wave groups can be recognized.

2. With point soundings at sea, because the separation between shot points is much greater than a wavelength (three to five times greater), correlation of individual phases is possible only in rare cases. As a consequence, groups of oscillations which have similar properties as individual events are identified and traced (see Fig. 4.1).

Thus, specific records obtained under various survey conditions require essentially unique procedures in the initial step of interpretation – the identification of waves. In continuous profiling with a high station density, the concept of wave groups is practically identical with the concept of individual events, while for segmentally continuous profiles, even when clear individual events are recorded, groups always include several individual waves. The length of a group varies accordingly. With segmentally continuous profiling, the lengths of groups recorded under various conditions are usually 1-3 sec, while for continuous profiling, the lengths may be reduced to 0.25-0.5 sec [69, 153, 154].

There are also significant differences in the ratio of velocities for a recognizable wave group. With a low-detail survey, it is possible to recognize wave groups in the late portion of the records which differ in velocity by at least 7-10% while with continuous profiling, it is possible to separate events which differ less in velocity (only by 3-5%).

These figures characterize only the averaged parameters for a group of deep waves for the various forms of profiling. Under actual conditions one rarely finds a consistent variation in the length or compactness of a group with distance. These are significant differences in seismic records from different tectonic provinces.

Figures 4.2-4.4 show examples of seismograms with compact and complex groups of waves.

For identifying wave groups and tracing them along a profile, as well as for determining the nature of a wave, the same velocity and wave form criteria are used as in the case of individual waves.

However, each of these characteristics does not have precisely the same physical meaning for a group of waves. Instead of having a travel-time curve for a phase of a wave, we have a travel-time curve which corresponds to the beginning of a wave group, that is, a travel-time curve for the arrival of the first wave in a group, which may shift from one phase to the next

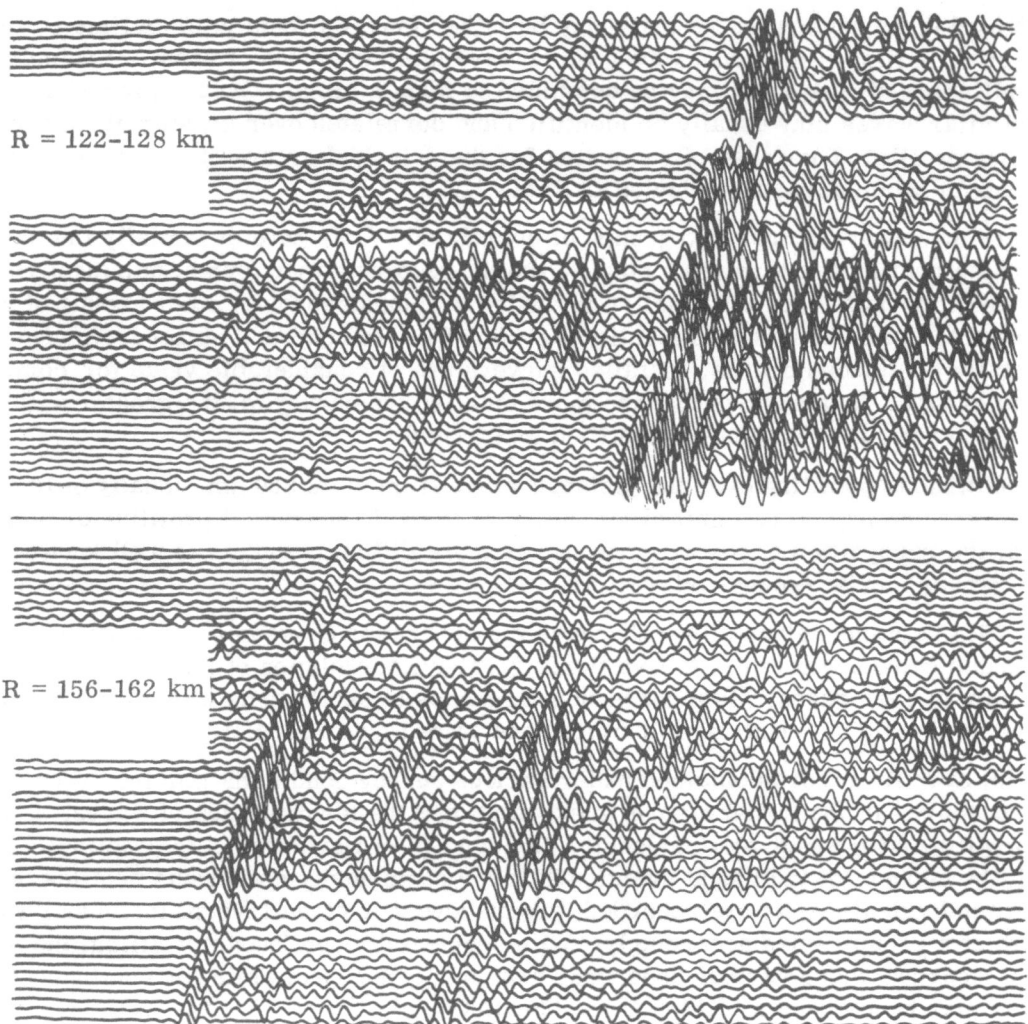

R = 122–128 km

R = 156–162 km

Fig. 4.2. Deep seismic sounding records for continuous profiling, compact wave groups, Ukraine, 1962 (Kiev Geophysical Trust, Demidenko and others [66]). Geophone spacing, 100 m.

along a profile. All other velocity characteristics of a group are determined in terms of this averaged relationship.

The travel-time curve for the beginning of a group, as we shall see, does not correspond to the travel-time curve for the maximum amplitude in the group, just as the travel-time curve for a beginning of a phase does not correspond with the travel-time curve for the maximum amplitude of the phase. The travel-time curves for later waves in a group may not be quite parallel with the travel-time curve for the first wave. In a further consideration of the correlation of wave groups, we will return to the basic idea of the correspondence between wave groups and individual waves. We should note in particular that if the first wave of a group corresponds to a single continuous smooth boundary, correlation of both phases and waves should be possible along a continuous profile. However, as has already been noted, in most areas such is not the case.

We will now formulate the physical basis which makes possible the combination of individual waves into groups, and the further interpretation of these as though they are single waves.

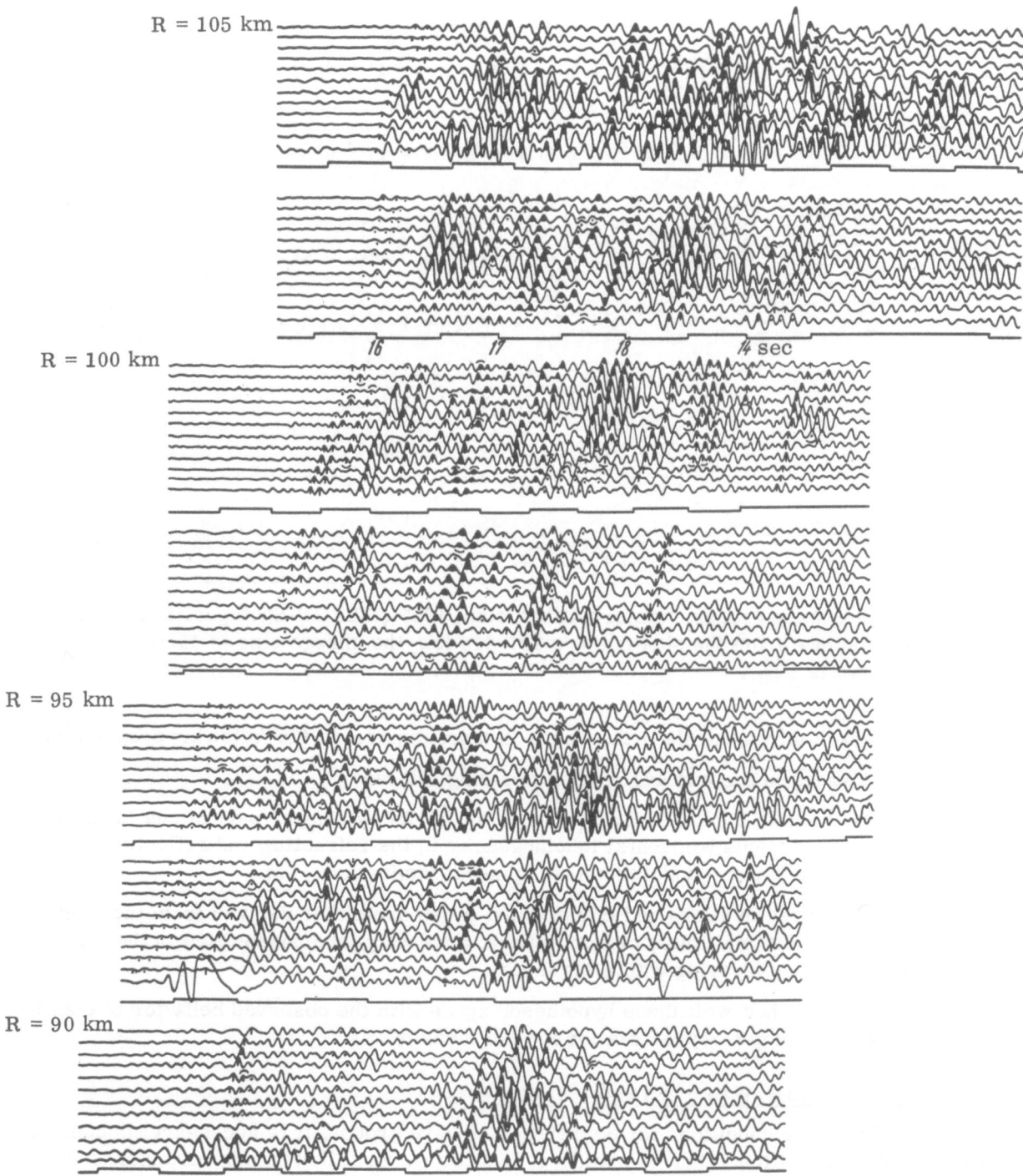

Fig. 4.3. Deep seismic sounding records, complex wave groups, Turkmenia, 1952 (Institute of Physics of the Earth, Kosminskaya and Tulina [95]). Geophone spacing, 200 m.

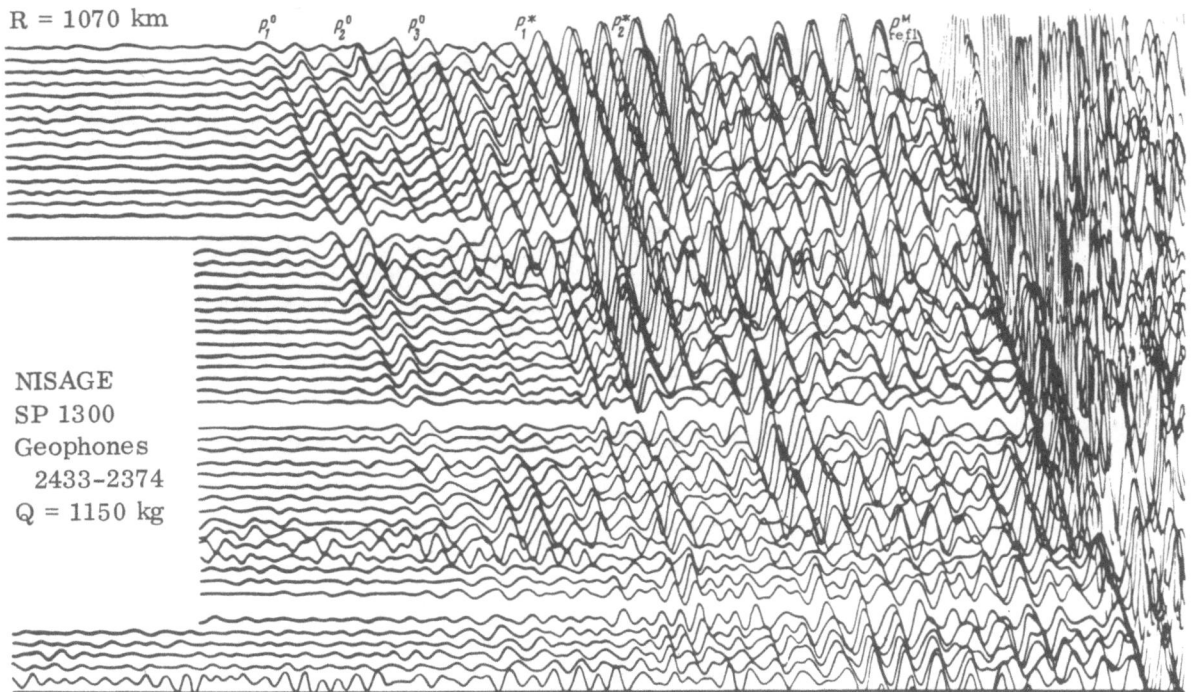

Fig. 4.4. Complex wave groups, western Uzbekistan, 1965 (NISAGE-Spetzgeofizika, V. Z. Ryaboi). The distance from the shot point is 107-113 km, the charge size is 1150 kg, and the dominant wave is P_{refl}^M.

This approach is reasonable if the waves in a group possess the following properties:

1. all the waves in a group are related to a single layer;

2. the first wave in a group corresponds to the surface or upper part of one layer; the shift of this wave represents horizontal inhomogeneity in the refracting and reflecting properties at the top of the layer — undulations, roughness, laminations, etc.;

3. the later waves in a long and complicated group are either derived from the first wave (dilatational waves, multiples) or correspond to velocity inhomogeneities in the form of irregular, discontinuous layering within the layer.

We will consider how well these hypotheses agree with the observed behavior of elastic waves and we will develop a representation for the medium.

§ 2. Nature of the Wave Field

In elucidating the conditions for correlation of wave groups, it is necessary, in considering records obtained with continuous profiling, to examine the following characteristics of wave propagation:

1. the correlatability of individual waves and wave groups;

2. the nature of phase and group travel-time curves; and

3. the intensities of the individual waves and wave groups and their variation with distance.

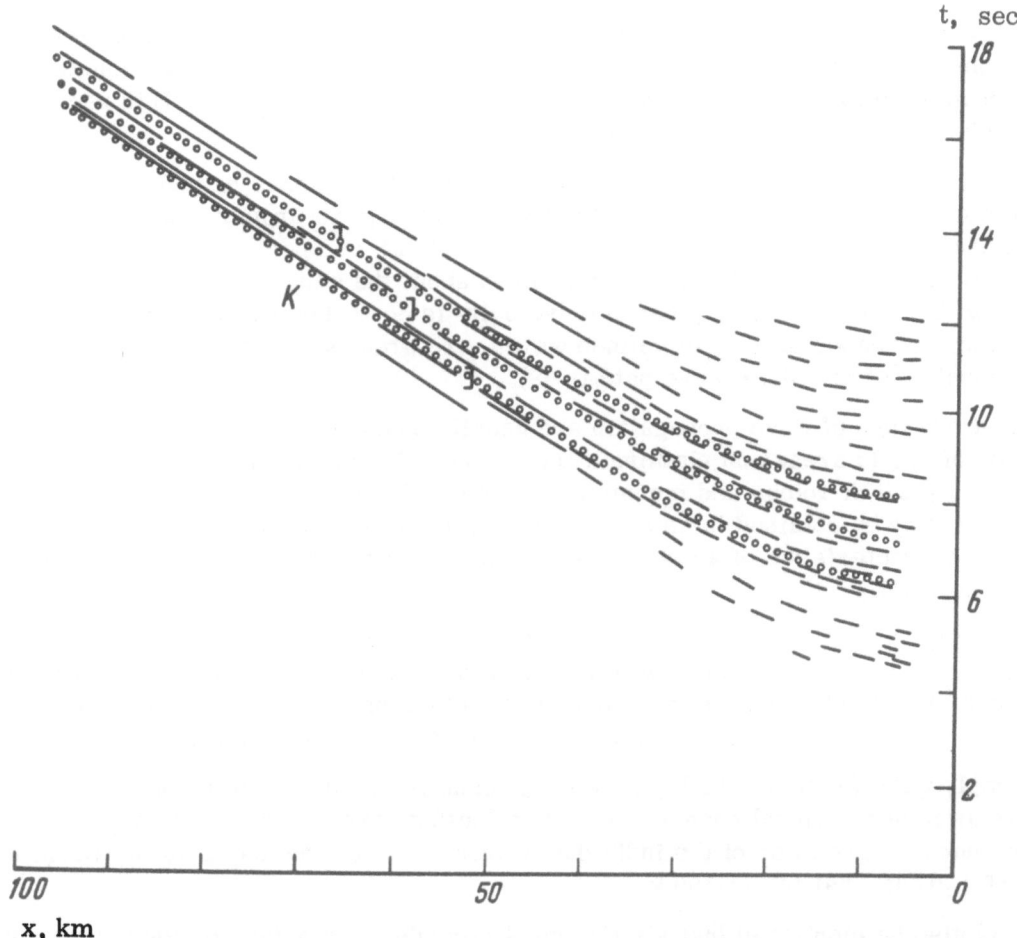

Fig. 4.5. Reflected waves at less-than-critical and more-than-critical angles, Ukraine. The circles are the theoretical travel-time curves for reflected waves. The vertical bars indicate the critical points (Chekunov [184]).

1. The Correlatability of Individual Waves.   Data on the correlatability of individual waves in various areas were given in Table 12.

The interval over which a wave may be correlated varies over wide limits, but only in rare cases does the average extent of interpretation exceed 20-30 km, which is usually no more than one-third of the range over which a wave group can be traced. The following consistencies in behavior have been observed: the correlation range for dominant waves is larger (amounting to 50-100 km) than for nondominant waves; as a wave becomes the first arrival, its correlatability generally improves; and in areas with complex tectonic zones, the range of correlation (including dominant waves and first arrivals) is less than in undisturbed platform areas (see Figs. 4.2 and 4.3). This general behavior has superimposed on it various specific aspects not only in different regions, but even along a single profile for records obtained from different shot points. These also differ for different wave groups. Thus, with the $P^M_{0\,refl}$ wave, there are rarely areas with only short intervals over which phases may be correlated, while the much weaker $P^K_{1,2}$ wave frequently may be traced over a longer interval, and conversely.

The waves with the best range for correlatability with continuous profiling belong to the first group. With increased distance from the shot point, waves with different apparent velocities group together and are traced as they interfere with each other. Because the various waves differ only slightly in velocity, the length of an interference zone can be considerable, and the recorded wave forms for the interfering waves can be quite clear. This permits correlation of individual phases in such a zone.

Individual waves, collected into groups in various tectonic zones have different properties. Three cases may be recognized, with three types of wave propagation. In the first, the middle wave of a group is strongly dominant, so that a short, compact group is formed (Fig. 4.2) [66]; in the second, the individual waves all have about the same amplitudes, and their identification with a particular group based on velocity is conditional [8]; and in the third, waves which despite an abundance of clearly defined waves, are grouped together because of an average "inhomogeneity" of the whole wave pattern [8].

These three types of wave propagation are usually characteristic of different areas: the first is typical of shield areas and platforms [113, 131] and some basins [42]; the second and third types are typical of folded and mountainous regions [8, 63, 96]. However, all three types of records may be obtained in an area with a uniform tectonic setting. Also, frequently, a change in the characteristics of a given group of waves or several groups may take place along a traverse [131].

In examining questions about the nature of layering in the crust, the case of clearly traceable dominant waves corresponding to intermediate boundaries in the crust is of particular interest. Highly characteristic records of this type have been obtained along a deep seismic sounding profile from Zvenigorodka to Novgorod-Severskiy (Fig. 4.2).

The simple wave form and the high intensity make it possible in this case to carry out a limited correlation of individual phases, without collecting several waves into a group. The extent of continuous correlation of the individual waves is nearly the same as the range for correlation of a group, that is, 80-100 km.

It should also be mentioned that similar good records were obtained along some small portions of the profile from Issyk-Kul' to Balkhash, though at that time interpretation was based on correlation of wave groups [50].

It should also be noted that records obtained in the area for critical-angle reflections have a unique form. In this case, individual waves usually have less extent, with the range of correlation not usually being more than 1-3 km, rarely being as much as 8-10 km. The intensity of these waves is frequently quite high, but, as has been observed, there is no consistent variation with time or distance for a combination of several waves which occur over some short-time interval.

Several investigations have been carried out [222] making use of statistical analyses to study the density of phase correlations as a function of time, and it was concluded that for this time interval, there is an increase in the density of correlations for individual phases. The nature of the changes in these intervals with distance and comparison with records made at large distances allow us to assume that the zone of high-density reflected waves near the shot point corresponds to groups of dominant high-angle reflections (Liebschier [222], Chekunov [184], Yaroshevskaya).[†]

[†] N. I. Davydova, A. N. Fursov, and G. A. Yaroshevskaya, "Description of attempts to record low-angle reflections with standard magnetic recording equipment in the Bukhara region, 1964-1965," Inst. Fiz. Zemli, Akad. Nauk SSSR, Uzbek. Geophysical Trust, Moscow (1966).

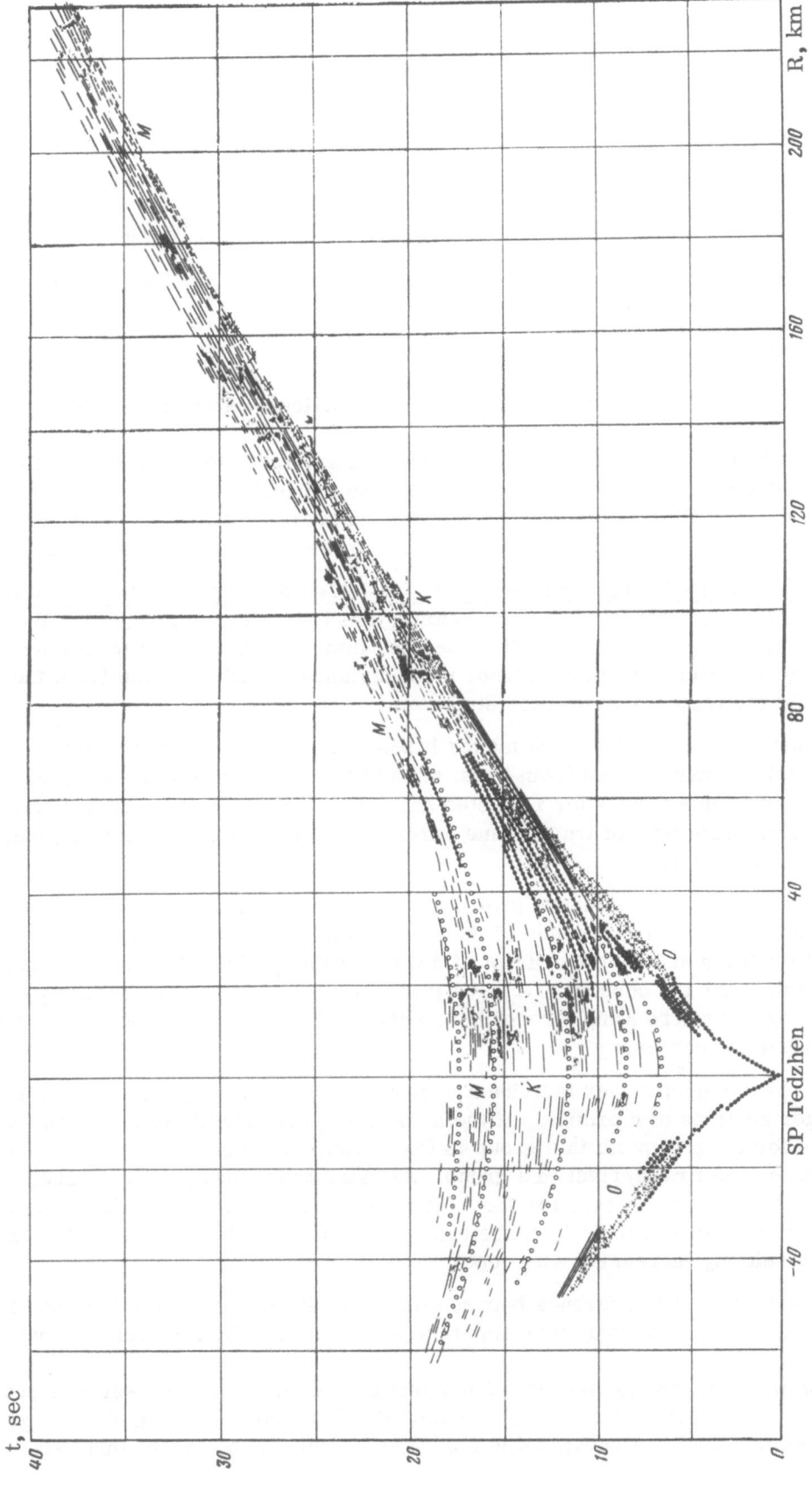

Fig. 4.6. Travel-time curves for reflected and refracted waves. Turkmenia (Belousov, Vol'vovskiĭ, et al. [20]).

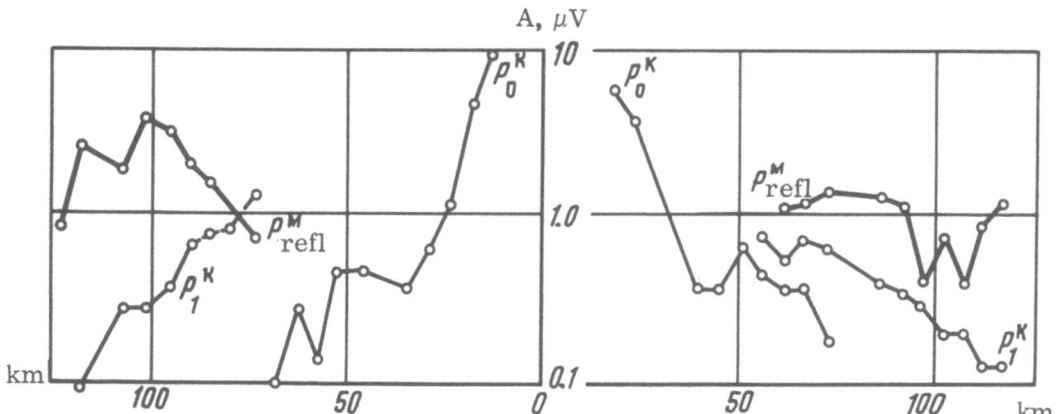

Fig. 4.7. Curves for the amplitudes of various waves [163, Chapter 7].

With all this, there are specific cases which are rare, in which clear deep reflections are recorded all the way from the shot point to the critical area (Baltic Shield area [114], Hungarian Uplight [233], Urals [179]). In these cases, the reflections either were dominant in amplitude or could be correlated reliably above the noise background along a profile.

2. Travel-Time Curves for Individual Waves or Groups of Waves. The general form of sets of travel-time curves has been shown in Figs. 4.5 and 4.6. The correlations between phases on adjacent traces are intermittent in form with a general tendency to fade away with time and with distance up to distances of 150-200 km from the shot point, and with the reverse tendency at greater distances.

Collecting waves into groups makes it possible to systematize the set of travel-time curves, which as may be seen from Figs. 4.5 and 4.6 corresponds to the typical form of a family of travel-time curves for reflected and refracted waves for a horizontally stratified medium. This same type of travel-time curve is also observed in seismological studies of the mantle (see Fig. 5.12).

With segmentally continuous profiling and point sounding, the grouped travel-time curves have a much simpler form because of the lesser number of wave groups which may be identified. With point profiling at sea, the concept of a wave group, as has already been indicated, coincides partially with the concept of a wave (see Fig. 4.1). Therefore, grouping of travel-time curves in regions where simple records are obtained practically coincides with the use of phase travel-time curves.

The phase velocity over individual segments is somewhat greater than the group velocity along longer sections of a profile. At the same time, the velocity determined from the travel-time curve for the maximum in a group is lower than the velocity determined from the travel-time curve for the first arrival in a group. As calculations have shown,[†] such a relationship between the travel-time curves for the first waves in a group and for the wave with the maximum amplitude is characteristic of the case of interference between waves in thinly laminated media with velocity increasing with depth.

The matter of the difference between the travel-time curves for individual phases and for the first arrival of a group may apparently be explained by the averaging which is done

[†] I. P. Kosminskaya, "Some aspects of geometrical seismics in deep seismic sounding" (Discussion by I. P. Kosminskaya, L. E. Aronov, G. G. Mikhota, and Yu. V. Tulina, Concerning Deep Seismic Sounding Operations in the Pamir-Altai Zone), Fondi. Inst. Fiz. Zemli (1957).

when apparent velocity is determined from grouped travel-time curves and by the effect of attenuation on the first waves in a group. With reliable tracing of individual waves along a profile, these travel-time curves are very close or coincide.

   3. The Intensity of Waves and Wave Groups. We will examine the intensity of individual waves and wave groups and the manner in which it varies with time and with distance from the shot point.

   We have already considered the general intensity levels and their variation with distances in Chapter III. As has been shown in papers by A. S. Alekseev and others [3, 5, 6, 163], the change in amplitude of reflected and refracted waves with distance is generally what one would expect in a horizontally stratified medium.

   We will dwell on the relationship between the intensity of individual waves and of wave groups. The individual waves in a group, as we have seen, may all be at about the same level, but they may also differ markedly in amplitude (Fig. 4.7). The form of a group may be characterized by the relative intensities. We distinguish between three types of grouping† : 1) the maximum corresponds to the first wave in a group and so is located at the beginning of the group; 2) the maximum corresponds to a later wave in the group; and 3) the grouping has no distinct maximum. Examples of such groups were shown in Fig. 4.1.

   A change in the form of a grouping with distance from the shot point indicates either a change in the properties of the individual waves in the group or in their mutual interference. A shift in the maximum toward the later part of a group for large transit times may indicate either a change in the relative intensity of the waves in the group or the effect of superposition on other waves or adjacent groups.

   The source of inconsistent changes in the intensity of individual waves in a group may be interference between groups. Therefore, the complexity of a group apparently will change with distance from the source, with the maximum of the group shifting toward later times; that is, there is a transition from the first type of grouping to the second or the third type. From this viewpoint, it is an easy matter to explain the observed differences between the slope of the travel-time curves for the first waves in a group and travel-time curves constructed on the basis of arrival times for the maximum in a group. The type of group will also be a function of the length taken for the group. With collection of events into groups with lengths greater than one second, the groups may have a complicated, irregular form, and the maximum in the group may shift with respect to the beginning of the group. In this case, because of grouping of several well-defined but shorter groups into a single group, an analysis of the interference effect might lead to a second type of grouping [36].

   The intensity of a wave group is determined from a curve A(R) which is constructed differently for the different survey systems. For point and segmentally continuous profiling the maximum amplitude in a group is used, while for continuous profiling the curve is constructed using the maximum phase of the first wave in the group.

   If it is assumed that there is systematic interference between the first waves in a group and the background noise, both methods for constructing the A(R) curve will lead to an erroneously flattened-out curve under the condition that each successive wave is more intense than the preceding.

---

† This classification was introduced in a program of generalization done by the Deep Seismic Sounding Group assigned to problems of integrated investigations of the crust and mantle, Department of Science of the Earth, AN SSSR.

The effect of flattening out of the A(R) curve may also be explained by failure to account adequately for the increase in charge size with increasing distance from the shot point. It is possible that the way in which data are reduced may explain the anomalously small attention of waves related to boundaries in the lower crust, such as is noted, for example, on the Russian platform (Pomerantseva [131]).

In comparing data on the dynamic characteristics of waves obtained with various survey systems and reduced by various means, it is necessary to keep in mind that they characterize essentially different properties of the medium. A curve for A(R) constructed on the basis of groupings of the early waves is determined by waves related to the surface layers. A curve for A(R) corresponding to a collection of all wave groups provides an idea about the variation in intensity of the entire wave assembly, and as a result, with wave groups of the second and third types, the curves for A(R) may differ significantly from the curve for A(R) for the first waves.

Examples of curves for A(R) are shown in Fig. 4.8. Their behavior is characterized by different rates of attenuation for the various waves, which is related basically to different propagation functions for refracted (head) and reflected waves from boundaries at various depths. The ratios of amplitudes are qualitatively in agreement with the uniform layered model, but there are a number of discrepancies with this model. For some waves, particularly the refracted waves, the best agreement between observed and computed characteristics of the field may be obtained with the inclusion of inhomogeneous layers in the theoretical model [3, 131, 153, 163, Chapter 4].

The disagreement between intensities of observed waves and theoretical calculations of the accepted models may be explained also by failure to consider the effects of interference, as noted earlier (this chapter, section 4).

4. Group Correlation in the Case of a Multiple Layered Crust. It is apparent from the examples of records and travel-time curves that have been presented that with continuous profiling, group correlation results in a systematization of a discontinuous wave field, expressing the average characteristics of the change in apparent wave speed with time and distance. These characteristics usually are well defined for various shot points along a profile. Further analysis of the wave field, based on determining the nature of the waves, is done using the system of grouped travel-time curves under the assumption that they represent individual waves.

This analysis indicates that a family of grouped travel-time curves contains high-angle reflections, low-angle reflections, and refracted waves that are similar to head waves and/or are clearly refracted [3, 5, 163].

The dynamic characteristics of a wave group (the wave forms) and their behavior (the ratios of intensities of various waves and the change with distance) agree with the travel times for waves traveling in a uniform layer or weakly inhomogeneous medium to a first approximation.

Consideration of travel times and wave forms makes it possible to determine the properties of the seismic section. The solution of the direct problem for a variety of models of such sections permits an evaluation of the effect of wave attenuation related to propagation, and, considering the limitations of the recording equipment, an indication that the recording range for various groups of deep waves corresponds to the relationship for an ideally elastic medium, which is evidence for the low attenuation in the crust. This is substantiated by the clear form of the records, particularly at low frequencies. All of these properties of wave groups, based on detailed investigations in many areas with different crustal structures (Alekseev et al. [5], Pomerantseva [131], and Aver'yanov and others [163, Chapter 4], as well as [3])

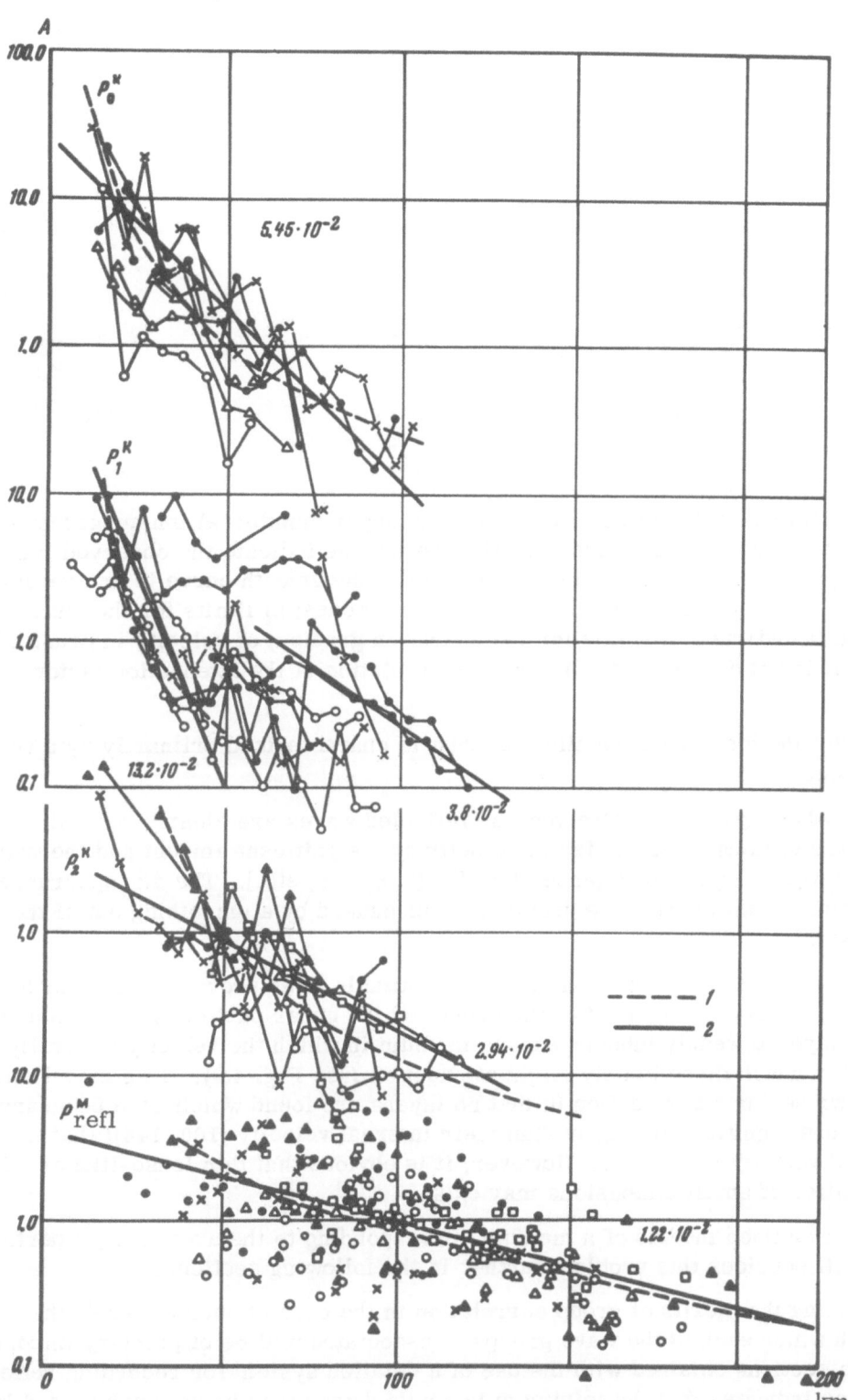

Fig. 4.8. Amplitudes of deep waves (Sea of Okhotsk [163]): 1) theoretical curves for head waves; 2) averaging lines used in calculating α; the various values indicate data from various locations. Figures to the right are effective attenuation coefficients, per km. Curves are based on maximum amplitudes in groups.

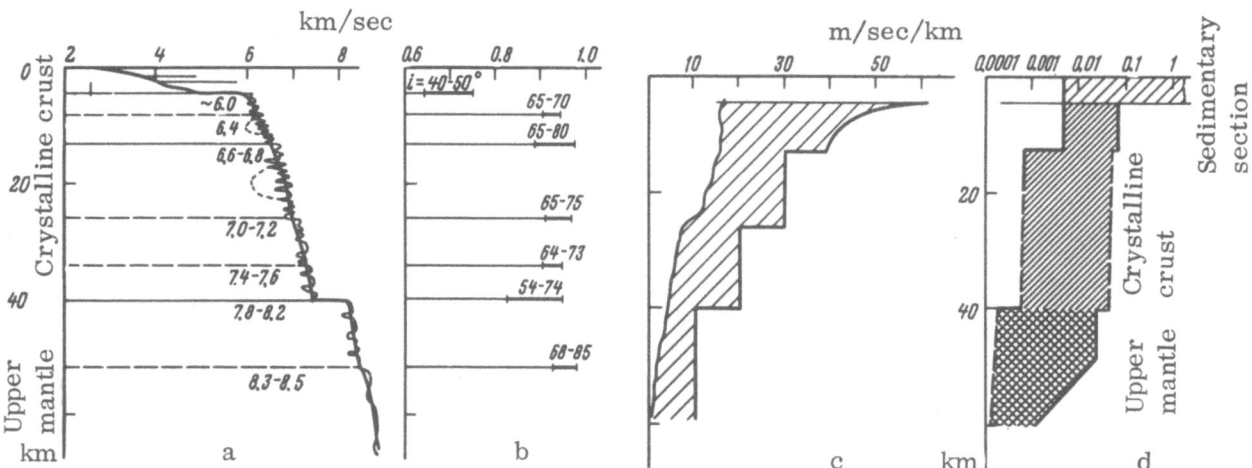

Fig. 4.9. Physical characteristics of the crust and upper mantle: a) inhomogeneous layered model of a continental crust and upper mantle; the figures indicate the observed limits for velocity; the wavy curve is an assumed velocity profile; the smooth curve is an averaged velocity profile; the dashed curves show possible low-velocity zones; b) limits for the change in relative velocity at boundaries; c) limits for the change in gradient of velocity in crustal layers (the lower limit is not determined); d) limits for variation of the attenuation factor.

do not contradict the idea that basically the crust is characterized primarily by a velocity as a function of depth.

In areas where groups of refracted and reflected waves are clearly defined, it may be assumed that the various layers differ in velocity by a significant amount and the transition in velocity from layer to layer takes place abruptly [3, 5, 116, etc.]. The disappearance of a single wave group or all of the wave groups may be caused by a smoothing out of these velocity discontinuities.

Thus, the deep seismic sounding data now available to us provide a reasonable basis for considering that the crystalline part of the crust and the upper part of the mantle may best be modeled as a layered weakly inhomogeneous medium in which the velocity generally increases with depth and in which there is only slight attenuation (see Fig. 4.9). This model differs from the better-known sedimentary section in that no layers are found which have boundary velocities which are significantly higher than their internal velocity [100, 144]; that is, there are no well-defined velocity inversions. However, it is obvious that thin lense-like or other forms of inhomogeneities of small dimensions may exist.

We have examined models of a medium corresponding to the averaged properties for wave groups. We will consider this problem further in the following sections.

In evaluating the merits of group correlation in the case of such a model, the thickness of the layers with which each of the wave groups is associated will be of primary importance. An analysis of the sections obtained with the use of a detailed system for recording reflected and refracted waves indicates that the minimum layer thickness may be as much as 4-6 km (Chapter III, §4). This means that for a depth of 30-40 km to the M-discontinuity, 4 to 6 boundaries may be found in the section. In group correlations under these conditions, a question about what effect the factors characterizing a laminated medium may have on the accuracy of determining group travel-time arises.

TABLE 13. Apparent Velocity (km/sec) of the First
Attenuated Waves in a Group

Initial velocity, km/sec	Apparent velocity, km/sec		Initial velocity, km/sec,	Apparent velocity, km/sec	
	δt, sec			δt, sec	
	0.25	0.5		0.25	0.5
6.0	5.82	5.65	7.1	6.86	6.65
6.6	6.41	6.20	8.2	7.92	7.60

δt is the shift in the apparent first arrival.

We will consider two questions in relation to this: the effect on the velocities and wave forms of attenuation of the first waves with distance and the nature of interference of two adjacent wave groups.

Attenuation of the First Waves in a Group. Let us find the change in the first waves with distance from the shot point which is characteristic of attenuation of the first phases. We will assume that over a profile interval of 50 km in length, we will not only shift from phase to phase, but we will lose one or even two waves. This is sometimes possible in the case of poorly expressed records of the first waves. This effect results in a lowered velocity for the first wave, as indicated by the figures in Table 13.

Interference Between Two Adjacent Wave Groups. A distortion in apparent velocity is quite probable in the case of combined records of two wave groups in the range over which they interfere and it is impossible to determine the first arrival for each of the groups.

For the range of interference between K and M waves, because the M-waves are dominant, this distortion results in an increased velocity, while following this zone, the effect will be different for the waves $K + M_{refl}$ and the waves $M_{0,1,2} + M_{0,1,2...refl}$ (see Fig. 3.6).

The amount of increase or decrease in the velocity V_Σ for a complicated wave will depend on the difference in amplitudes of the interfering waves. For equal amplitudes, the phase velocity V_Σ for the complex oscillation will be the geometric average of the velocities V_1 and V_2 of the constituent waves (Kosminskaya [90]):

$$V_\Sigma = \frac{2V_1 V_2}{V_1 + V_2},$$

while in the case in which one of the waves is dominant, it will be given by the expression

$$V_\Sigma = V_2 \frac{\eta + 1}{k + \eta}, \tag{4.1}$$

where $k = V_2/V_1$ and $\eta = A_2/A_1$; A_1 and A_2 being the amplitudes of the constituent waves. Formula (4.1) is the most appropriate for good correlation of complex oscillations — the range for inphase or nearly inphase superposition of the waves.

Velocities of interfering waves are given in Table 14 for the case of superposition of adjacent (with respect to travel time) wave groups for the section in western Turkmenia (Egorkin [69]).

The nature of the propagation effects for strong reflected and refracted waves with distance for a horizontally layered model of the crust and mantle is such that within an interval of 50-200 km each

TABLE 14. Velocity of Interfering Waves

Velocity of interfering waves, km/sec		Phase velocity of wave complex, km/sec		
		η		
V_1	V_2	1	3	5
6.1	6.3	6.15	6.22	6.25
6.6	6.8	6.70	6.72	6.75
7.1	7.3	7.20	7.25	7.27
7.6	8.2	7.90	8.05	9.10
8.2	8.6	8.30	8.50	8.53

TABLE 15. Velocity of the First Wave in a Group with Consideration of Attenuation and Interference

Velocity of the first wave in a group, km/sec	Velocity, in km/sec, of the first wave with attenuation and interference being considered, $\eta = 3$, L = 50 km		Velocity of the first wave in a group, km/sec	Velocity, in km/sec, of the first wave with attenuation and interference being considered, $\eta = 3$, L = 50 km	
	$\delta t = 0.25$ sec	$\delta t = 0.5$ sec		$\delta t = 0.25$ sec	$\delta t = 0.5$ sec
6.0	6.02	5.93	7.1	7.05	6.92
6.6	6.56	6.46	8.2	8.22	8.01

L is the distance over which a wave can be traced.

successive wave will be more intense than the preceding, i.e., η will be greater than unity.[†] In this respect the results of computations done with $\eta = 1$, 3, and 5 are given in Table 14. These data show the effect of interference may be to increase velocity (with respect to the velocity of the earliest waves) by 0.1–0.3 km/sec for equal amplitudes, $\eta = 1$, and by 0.2–0.4 km/sec for $\eta \geq 3$.

Comparison of Tables 13 and 14 makes it possible to evaluate the error in velocity when waves are attenuated or when successive waves interfere. For $\delta t < 0.5$ sec and $\eta \geq 3$, these two effects offset each other. For $\delta t > 0.5$ sec, which is possible, in particular, for segmentally continuous or point surveys, attenuation has the larger effect and the velocity will always be lowered somewhat, despite the effect of interference by nearly 0.1–0.2 km/sec (Table 15).

At distances of more than 200 km, that is, greater than the distance at which the mantle wave P_0^M emerges as first arrival, two different situations may occur. For the $P_{0,1,2}^M$ wave, the relative velocities will be the same as those computed, but for the crustal waves, which are traced in the later parts of the record (including the $P_{0\,refl}^M$ wave), an inverse relationship may be found; because in the case of the superposition of two reflected waves, interference will cause a lowering of the velocity with respect to the velocity of the first waves, which have

---

[†] Computed curves for the intensity of reflected and refracted waves based on equations from Petrashenya and others [43] have been prepared by a large group of authors for nearly forty uniformly layered crustal models of various types and about 20 inhomogeneous layered models. See, for example, N. I. Davydova, E. N. Zaitseva, and I. P. Kosminskaya, "Album of curves for the intensity of longitudinal waves for various crustal models," Fondi Inst. Fiz. Zemli, Akad. Nauk SSSR (1964).

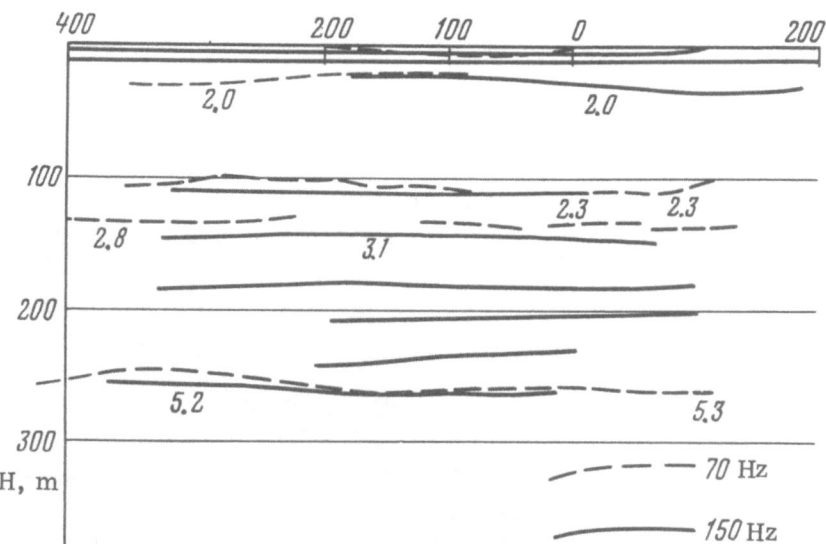

Fig. 4.10. A section obtained in seismic exploration in which medium and high frequencies were recorded (Berzon [23]). The figures are boundary velocities, km/sec.

amplitudes greater than those of later waves [in (4.1) this corresponds to k > 1 and η > 1, or V_Σ < V_2].

We have not considered here the case of interference near the critical point, but this will be discussed later.

Nature of the Identification of Wave Groups for the Study of a Multiple Layered Medium with Various Frequencies. From the geometric point of view, collection of waves into groups is equivalent to reducing the recorded frequencies. This has been well illustrated by examples from seismic exploration experiments (Berzon [23]).

Figure 4.10 shows a section constructed from data recorded at significantly different frequencies (70 and 150 Hz). A comparison of the two sets of boundaries indicates that in addition to all of the boundaries seen at low frequencies, a number of new boundaries may be traced at the higher frequency.

We have found good agreement in the locations of specific boundaries with interpretation of records for different frequencies, and the values for boundary velocities are practically the same. However, it has been stressed in [23] that boundaries have been constructed on the basis of travel-time curves for first arrivals. At the lower frequencies, the travel-time curves for phases in complex waves in some cases had a somewhat higher velocity than that for individual waves from the same interfaces, but recorded with higher-frequency filter settings.

We will examine a similar example of a finely laminated model of the crust. Two sets of travel-time curves are shown in Fig. 4.11 (Egorkin [69]); the first one was prepared from records obtained with continuous profiling, while the second (P_0^K, P_1^K, P_0^M) was based on groupings A(t) constructed with the same seismograms. The velocity profiles for the corresponding travel-time curves and for the first arrivals of P_0^K and P_0^M are given in Fig. 4.12.

A comparison of the velocity boundaries indicates that we could identify the boundaries d_0^K and d_0^M with reasonable accuracy with group correlation of the second series of travel-time curves. The intermediate boundary d_1^K is recognized at depths which do not correspond to the values of velocity in section 1.

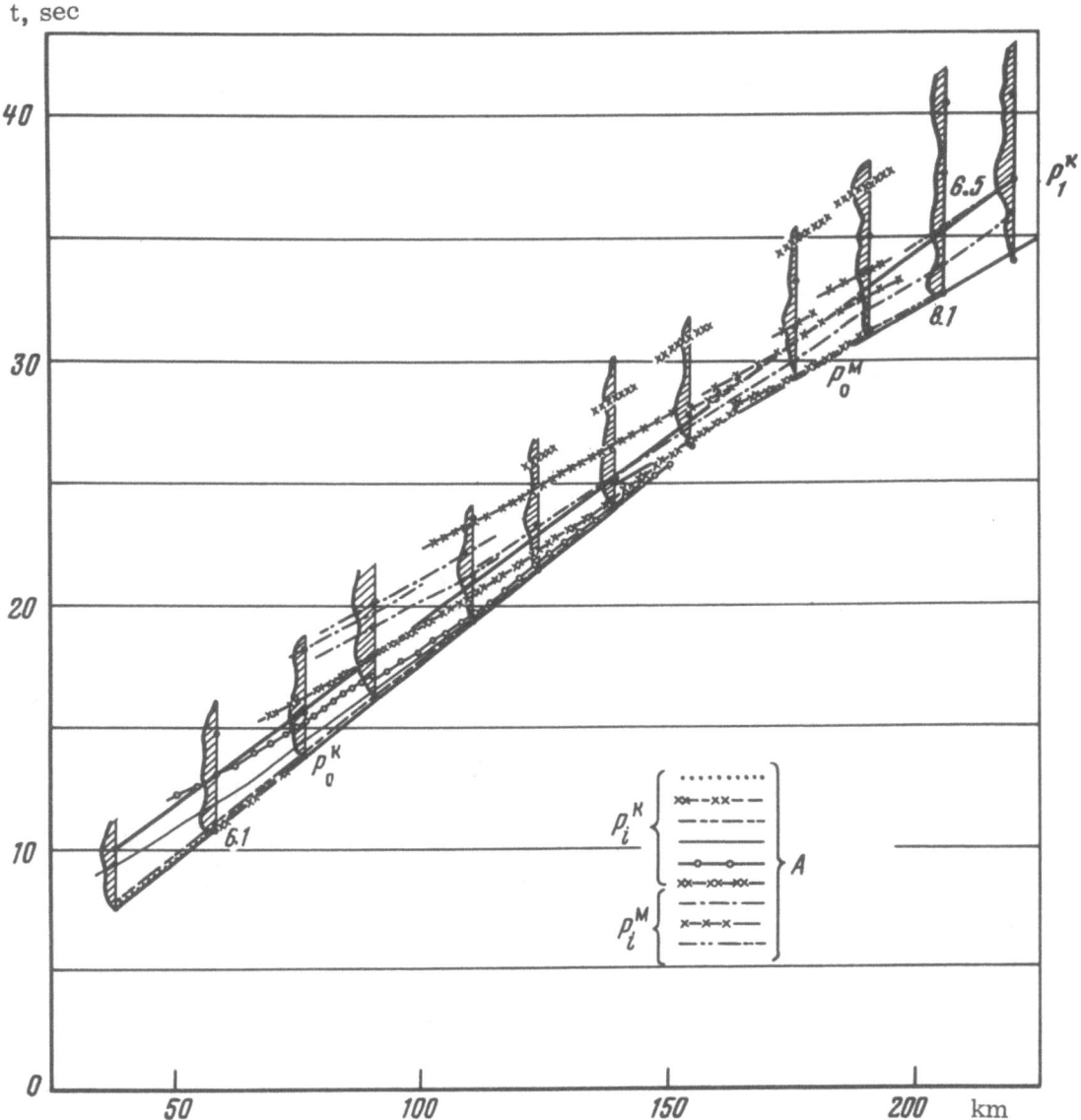

Fig. 4.11. Deep seismic sounding travel-time curves, constructed for the first waves in groups (fine lines) and the maxima in groups (heavy lines). Turkmenia, Karashor—Kara Bogaz Gol profile (Egorkin [69]). The shaded areas indicate the grouping of the seismic records; figures are velocities, km/sec.

Still greater differences between the sections are seen for version 3, constructed on the basis of travel times for the first arrivals, and which corresponds essentially to the seismological case (that is, a frequency of about 1 Hz) for which the crustal waves K in the later arrivals are not divided into groups, as indicated in Fig. 5.12.

The causes of the differences between the various versions of the sections are:

1. differences in the depth to the boundary d_0^M, constructed on the basis of the first waves, are related to differences in velocities above the M boundary, which is determined separately for each set of travel-time curves. All of the details of the velocity section, including the

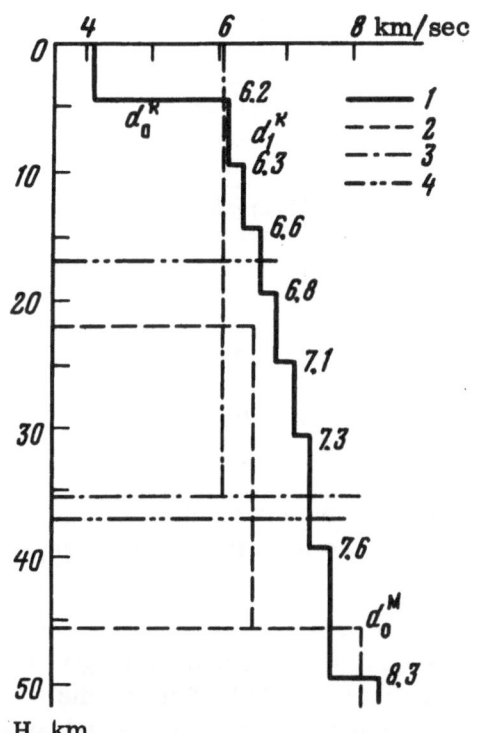

Fig. 4.12. Velocity sections for the travel times in Fig. 4.11. Curves: 1) first waves in groups (Egorkin [69]); 2) the waves P_0^K, P_1^K, and P_0^M; 3) the waves P_0^K and P_0^M; 4) boundaries found in this region with point soundings (Donabedov [68]).

velocity in the underlying layers (those with a velocity of 7.1-7.3 km/sec) are determined from the travel-time curves for the first waves in groups (Fig. 4.11). These layers are not resolved using travel times for P_0^K, P_1^K, and P_0^M, and with the use only of the first arrivals (travel times for P_0^K and P_0^M), the lowermost layer is one with a velocity of 6.5 km/sec, depending upon the way waves are grouped. The average velocities in the crystalline crust are 6.60, 6.31, and 6.05 km/sec, respectively, for the three versions;

2. the large depth for the intermediate boundary d_1^K on section 2 for small velocities V_H = 6.5 km/sec is related to the use of travel-time curves constructed using maxima in groups which, as we have seen earlier, always reduce velocity and increase time. The best agreement may be obtained by using the beginning point of a wave group, as has been recommended in many references [36, 96], which permits an approximation to the first arrival of the intermediate group, which is easily recognized by wave form, when surveys are made with the segmentally continuous and point sounding techniques. This makes possible some improvement of the results and an approximation to section 1 for the boundary with a velocity of 6.6 km/sec as has been done in [68].

However, with poorly detailed data, we will always be forced to combine several groups of travel-time curves in series A into a single grouped travel-time curve of series B, as has been done in Fig. 4.1. In so doing, errors arise from the effects of attenuation and interference which are larger than those which occur when short wave groups are used in continuous profiling (Tables 13 and 14) and which may cause not only distortion of depths but also distortion of the form of a boundary.

Thus, in the case of a thinly layered medium, group correlation using a system with coarse detail results in a schematic section with two forms of error being present:

1. reduction in the depth to boundaries constructed from travel times for the first waves in groups, and;

2. distortion of the velocity section of the crust, related to the recognition of intermediate boundaries, the position of which and the velocity for which correspond only approximately to the detailed velocity section. The boundary velocity for intermediate boundaries is characteristic of a minimum velocity in the layer beneath the boundary, which may increase significantly with depth.

Thus, in surveys made with the continuous profiling technique, many groups of deep reflected and refracted waves having apparent velocities and wave forms corresponding to a multiple layered piecewise homogeneous (or weakly inhomogeneous) medium are recognized. The propagation speeds and layer thicknesses in this medium are such that the schematic presentation of data lacking detail and obtained at the lower frequencies must lead not to true

TABLE 16. Depths to the M-Discontinuity for the Eastern Shore of the Caspian Sea, Determined with Techniques Providing Different Degrees of Detail

	Point of profiling		Continuous profiling: Karashor—Kara Bogaz Gol Profile [68, 69]	
	near-shore zone [3]	Bekdash—Karshi profile [68]	from reflections	from refraction velocities
Average velocity in crust, km/sec	5.5	6.1	6.2	6.4
Velocity for M-discontinuity, km/sec	8.1	−†	−†	8.3
Depth to M-discontinuity, km	30-32	34-36	35-37	49

† The M-discontinuity is determined from reflected waves.

average velocities for the section but to consistently low velocities in the crust and mantle, which in turn results in a consistent underestimation of the depth to the principal boundaries, and in particular, the M-discontinuity.

This conclusion is in good agreement with the observed lack of correspondence between velocity sections obtained along deep seismic sounding profiles carried out in neighboring regions using field techniques with differing degrees of detail [20, 42, 55, 68, 69]. Data on the average velocities and the depths to the M-discontinuity obtained from point soundings and continuous profiling in closely neighboring regions along the eastern shores of the Caspian Sea are listed in Table 16.

Two columns of figures are given for the profile Karashor—Kara Bogaz Gol. The first corresponds to a rough interpretation and construction of the M-discontinuity on the basis of strong reflections related to the surface of a layer with a boundary velocity of 7.6 km/sec (see Fig. 4.2). The second is based on a careful study of the velocity section and recognition of the M-discontinuity from refraction velocities. In this case, the M-discontinuity is defined as the upper surface of a layer with a velocity of 8.3 km/sec. This example illustrates the amount of scatter to velocities assigned to the crust which may exist when data are obtained with field techniques providing differing degrees of detail.

This scatter raises questions as to which of the boundaries is the M-discontinuity, and how its position is determined from the combined analysis of refraction and reflection data. Usually, in interpreting surveys which lack detail, waves from the lower crust which have a velocity higher than 7.5 km/sec cannot be identified, as they are combined with mantle waves, resulting in a reduction of the depth to the boundary of a layer with a velocity greater than 8 km/sec.

This same relationship is usually found in comparing deep seismic sounding data with seismological data. The depth to the M-discontinuity on the basis of seismological data is always less than that obtained on the basis of deep seismic sounding data (Pamir—Alyai [96, 110], Kazakhstan‡). This is explained, apparently, by differences in the information about the crustal velocity profile (in treating seismological data, no velocity greater than 7 km/sec is considered to be present). The M-wave in seismology corresponds to a boundary with a velocity of 8 km/sec, for which a refracted wave is traced as the first arrival, while in the case of deep seismic sounding, because of the weak signals and the greater attenuation of the high

‡ Personal communication, I. L. Nersesova.

frequencies in the first waves, later waves corresponding to deeper zones in the mantle are traced as the apparent first arrival. This may also be an explanation for the observed disagreement between the travel times for the first waves in deep seismic sounding and in seismological studies; the travel times from earthquakes are frequently less than those in deep seismic sounding. Another cause of the disagreement may be the different depths for the sources.

We have still to discuss the question as to what criteria may be used in deep seismic sounding sections to identify the M-discontinuity and this will be discussed in the following paragraphs.

In discussing the study of a thinly layered crust at various frequencies and with survey procedures which provide varying degrees of detail, we have not touched upon the physical properties of the boundaries which are seen, and the dependence of the corresponding waves on frequency. Some experimental data are available both from deep seismic sounding and particularly from seismic exploration experiments [23, 25] from which it is possible to conclude that the refracting and reflecting properties of even continuous (in the geological sense) boundaries are significantly dependent on frequency, particularly for a thinly layered medium. Keeping in mind the weak velocity contrasts in the crust, we may assume that at the different frequencies, different combinations of layer velocities or inhomogeneities may arise which are close on the average either to a homogeneous or an inhomogeneous medium, and which change along a profile. Such effects have already been used to explain observed frequencies (Chapter II, § 5). In such considerations, the study of the reflecting characteristics of a boundary in the simple case of near-normal incidence is of particular interest.

Thus, our concept that the use of grouped correlation is analogous to reducing the frequency is valid only as a first, geometrical approximation. It can be shown to be untenable with respect to the wave form characteristics of a wave group (reflection coefficients, attenuation spectrum) even in the case that the speeds of propagation are practically identical at the various frequencies. Such disagreement is particularly obvious for segmentally continuous profiling, for which the amplitude characteristics of waves and groups of waves correspond to a recorded frequency of about 10 Hz while it is possible to trace only groups of waves with a length that corresponds to the resolution obtainable at a frequency of about 5 Hz (see Chapter III, § 4). The wave forms which we have described for these travel-time curves may be significantly different from those which would be obtained if the observations had actually been made at a lower frequency, corresponding to the schematic travel-time curves.

The importance of this concept is confirmed by the correspondence between the frequency of the recorded vibrations and the field technique used, the positive tracing of individual strong events or compact groups of vibrations, and forces us to reflect on the fundamental need for an experimental study of the characteristics of deep boundaries using a variety of frequencies from explosions and earthquakes.

§ 3. Types of Waves

The question of determining the type of wave has received considerable attention. A fundamental investigation of the nature of waves and crustal models using elastic wave propagation theory has been carried out by A. S. Alekseev. He has published a number of papers, done in conjunction with his colleagues [4, 5], and reviewed from a general point of view in [6].

The wave form characteristics and travel times used in identifying the nature of waves in deep seismic sounding are the same as those used in seismic exploration. Therefore, I will review only briefly the important points used in identifying waves as reflected or refracted, head or surface. Because of the nature of some of the physical properties of the crust — the weak velocity contrasts, the low attenuation, and the complex blocky structure — these questions involve some difficulty and in order to obtain a reasonably specific solution, it is necessary to lay out the observation system appropriately.

   1. Reflection and Refraction of Waves at a Single Boundary. Reflected waves are recorded in deep seismic sounding mainly at greater than critical angles, and therefore, the travel-time curves are nearly flat over long intervals. This usually prohibits determining the nature of the change in apparent velocity V* with distance R with a reasonable accuracy such as should be obtainable from travel-time curves for reflections when $dV*/dR < 0$ and $d^2V*/dR^2 > 0$ [5].

   An important characteristic in recognizing that a wave is reflected is nonparallelism in travel-time curves, so that at greater distances from the shot point, the velocity for a curve segment becomes lower. However, at great distances, the successive travel-time curves are practically parallel. Therefore, statistical studies of the decrease in velocity with distance from the shot point may be used with a series of travel-time curves in which the apparent velocity changes markedly with distance, as has been done in reducing marine deep seismic sounding data [163, Chapter 7].

   Reflected and refracted waves from a single boundary are better distinguished on the basis of wave form.

   Reflected waves are usually dominant on the records. For some wave groups, and especially for the $P^M_{0\,refl}$ event, a maximum is observed on the amplitude curve in the region of the critical angle [163, Chapter 7].

   In the region of the critical angle, as has been pointed out in §4, there is practically no difference in arrival times for the refracted and reflected waves. At large distances, the reflections are usually weak, and it is rarely possible to recognize them in the background of seismic noise and other arrivals. Refracted waves (either head waves or geometrically refracted waves) are recognized as first arrivals, but many of these are apparently combinations of refracted and reflected waves, as a result of some interference phenomenon, despite the fact that their travel-time curves are nearly straight lines (see Fig. 3.6). The travel-time curve for the first arrival, if it can be definitely identified, corresponds to the travel-time curve for a geometrically refracted wave, according to calculations by Smirnova [158] and by G. S. Pod"yapol'skii.†

   Ratio of Amplitudes of Reflected and Refracted Waves. In areas where reflections are recognized as being the later arrivals, they are usually more intense than the refracted waves from the same boundaries. However, it may be shown that the ratio of amplitudes is lower theoretically for a boundary with a velocity discontinuity. In such cases (as in Fig. 4.13), it is assumed that the refracted waves are not weak head waves but intense geometrically refracted waves.

   However in many regions where high-angle reflections are observed, the first waves may be so weak that they cannot be recognized in the noise. In such cases, they may have propagated essentially as head waves; that is, the boundary to which they are related is sharp and defines a region with very nearly constant velocity beneath.

   Both of these situations are observed in regions with a continental crust and apply to reflections from the M-discontinuity or boundaries in the lower part of the crust.

---

† G. S. Pod''yapol'skii (Institute of Physics of the Earth, Academy of Sciences of the USSR) developed a program for computing synthetic seismograms for a four-layer medium in 1963/64.

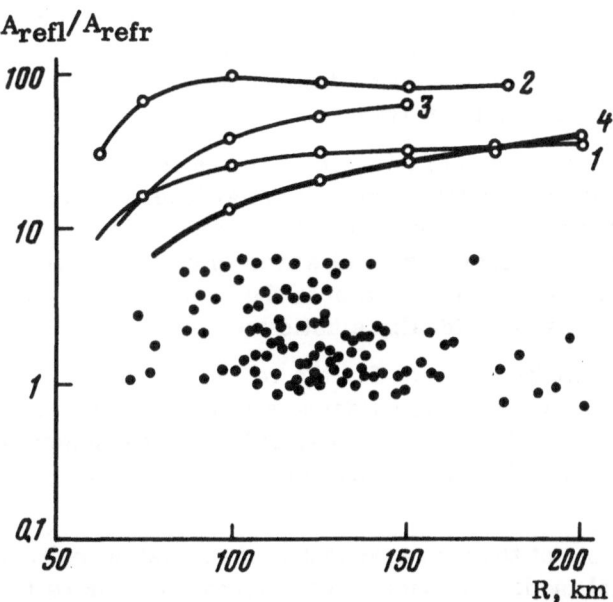

Fig. 4.13. Computed curves for the ratio of amplitudes for reflected and refracted waves for various types of crustal sections: 1) oceanic; 2) intermediate; 3) continental. The points are observed ratios for the northern and central parts of the Sea of Okhotsk ([163], Chapter 4).

Reflections from the M-discontinuity are not usually found in areas with an oceanic crust [102, 163, 170].

There is a considerable difficulty, which has not yet been resolved, in separating reflected waves which appear as later arrivals and refracted waves from the same boundary. These later waves are usually difficult to distinguish from other refracted waves appearing as first arrivals in other groups. Experimental work done along these lines by A. M. Epinat'eva and E. V. Karus [74] in seismic exploration has shown that with the presence of thin high-velocity layers in an alternating sequence with other beds, the refracted waves which we ordinarily assume to be head waves are actually interference waves formed by the superposition of several high-angle reflections from the boundaries in the sequence.

Apparently a similar effect takes place with crustal waves, but recognition of this effect when group correlation is being used is not possible.

It is important only to remember that intense waves, even if they are first arrivals in a group, are not ideal head waves.

2. Types of Refracted Waves. During the period 1957–1960 one of the primary questions in deep seismic sounding was that of the nature of refracted waves: are they head waves (boundary waves) or are they geometrically refracted waves (associated with a gradual increase in velocity with depth)?

The difficulty in distinguishing between these two types of waves, as also was the case with reflected waves, was related to the fact that because deep boundaries might be curved and velocity might change along a profile, it would not be possible to recognize the diagnostic condition for refraction using a single travel-time curve, i.e., $dV^*/dR > 0$ with $d^2V^*/dR^2 > 0$ (for the travel-time curve for a reflected wave, $dV^*/dR < 0$ with $d^2V^*/dR^2 > 0$).

Comparison of travel-time curves obtained for successive shot points along a profile for considerable distances (~50 km) between shot points and comparison of portions approaching the critical point in most cases do not make it possible to establish precisely the properties which are characteristic of refracted waves.

An analysis of the wave forms of some deep waves from a number of regions has indicated the possible existence of refractions in crustal layers (Aver'yanov [163], Chapter 4). It has been discovered that nearly universally, deep waves attenuate no more rapidly with distance and usually much less rapidly than head waves in a homogeneous layered medium. Moreover, if it is assumed that there is attenuation in the crust, head waves must be attenuated more rapidly than in the case of an ideally elastic medium.

Now there is also velocity evidence for the refractive nature of the P_0^K wave corresponding to the basement surface. This evidence was provided from a statistical analysis of a series of travel-time curves for three-, five-, and even seven-member intersecting groups. The nonparallelism of the travel-time curves amount to 0.1-0.2 sec per 50 km (Pomerantseva [131]).

A theoretical analysis of the characteristics of geometric refraction for an inhomogeneously layered crustal model (Alekseev, Aver'yanov, and others [4, 5, 163, Chapter 4]) indicates that the primary characteristics of geometric refraction in a medium are weak attenuation of the wave, the existence of multiple waves, and a limited range for recording. All of these features may vary, depending upon the nature of the section being studied. Thus, for example, such an important indicator as the limited range for recording is observed only for the P_0^K wave for the conditions of a near-oceanic crust. In the case of a continental crust, this zone is usually confused by the attenuation of waves with distance and interference.

These data on the characteristics of refracted and reflected waves are based on the velocities and wave forms for averaged properties of wave groups, traced in various regions. In an analysis of observed sets of reversed travel-time curves and travel-time curves from successive shot points along a profile, it is observed that the factor for each specific area which most affects the velocity characteristics for wave groups is relief on the deep boundaries (Kosminskaya [103]).

Therefore, in constructing a section from deep seismic sounding data, the first step is to neglect the possibility of geometric refraction and consider the refracted waves to be head waves. In cases in which there are horizontal boundaries for which it is possible to determine the vertical velocity gradient, appropriate corrections are made to the section (Puzyrev [136], Pomerantseva [131], and Aver'yanov [163, Chapter 4]).

This procedure makes it possible to determine some of the averaged physical properties of the crustal section in a given area from deep seismic sounding data: the velocity profile, the average gradient curve, and the average attenuation factor in layers. A seismic section for boundaries in crust and the upper mantle is constructed from the travel-time curves.

The physical properties for a continental crust have been indicated in Fig. 4.9.

The ranges over which the velocity gradient and the attenuation factor vary in the deep layers are indicated schematically in Fig. 4.9c and 4.9d. The gradient in the upper part of the crust amounts to 50-60 m/sec/km, and in the mantle, probably it is no more than 10 m/sec/km on the average. The attenuation factors have been computed on the basis of apparent attenuation coefficients. If, as has been indicated earlier, it is more probable from the wave form point of view that reflections and refractions exist for which the propagation exponent is less than 2, the apparent attenuation factor will apparently be of the same order as the true value [163, Chapter 4].

In comparing the conditions in deep seismic sounding with those in seismic exploration on Fig. 4.9, it is also indicated that maximal values for the attenuation factor apply for the sedimentary sequence. These values are one to two orders of magnitude larger than the average attenuation factor in the crust. Hence, it may be concluded that the wave forms for deep waves may also be more diagnostic than those for waves propagating through the sedimentary sequence. However, as we have seen, under actual conditions at frequencies around 10 Hz this is not found, a factor which is apparently explained by nonideal and nonhomogeneous deep boundaries.

The velocity contrasts in the medium are shown in Fig. 4.9a and 4.9b. The velocity ratios at some deep boundaries are larger than 0.98. Such boundaries have been but little studied, either experimentally or theoretically. The question as to the formation of reflected and refracted waves is particularly complex for such boundaries, either for normal incidence or for critical angles. The role of weak boundaries and layers with low velocities in the formation of the various body waves is not clear.

All of this complicates the analysis of deep seismic sounding data even for schematically grouped wave characteristics.

In this respect, we will consider several factors involved in the decay of wave amplitudes in deep seismic sounding: the power n involved in geometric spreading of a wave and the attenuation factor α for a wave (similarly, see Aver'yanov et al. [163, Chapter 4]).

3. Determination of n and α. As we have seen, it is necessary to record waves in the vicinity of the critical angle in order to observe the characteristics of the refracted and reflected waves in deep seismic sounding. In many cases, the observation procedure is incomplete.

In this respect, it is not possible to use directly the methods developed in seismic exploration for determining n and α in deep seismic sounding. This is related to two principal causes: 1) reflected and refracted waves are best identified in the vicinity of the critical angle, where the assumption that n = const is not valid; and 2) as a result of the nonuniformity of the medium along a profile, and the usual inadequacy of survey systems, attenuation is determined from a single curve rather than from a system of curves.

A third important factor which complicates the determination of α is the very small magnitude of this quantity, which is of the order of 10^{-3}-10^{-2} per kilometer, so that very accurate data are required with the exclusion of such random and unpredictable factors as geophone plant conditions, consideration of differing shooting efficiencies, consideration of local anomalies in the structure of the upper part of the section, and so on.

With variable conditions, the method used in averaging data for observations at each location on a land profile has a very large effect on the form of the A(R) curve. However, even in marine operations, which are largely free from these problems, the A(R) curve is not particularly smooth.

As a result, the procedure most commonly used in deep seismic sounding for determining the parameters α and n is as follows.

Initially, all of the observed curves are analyzed and those which correspond to horizontal segments of interfaces are selected. Then, the value for n is determined by plotting these curves to a bilogarithmic scale. For curves with n > 2, the value for α is determined by subtracting a curve A(R) calculated for the given crustal section which is determined from reduction of the travel-time curves from the observed data.

For amplitude curves of refracted waves with n < 2, an effective factor α_{eff} which will be essentially equal to the true value for refracted waves in the case of the weak gradients which are typical for the crust, is determined.

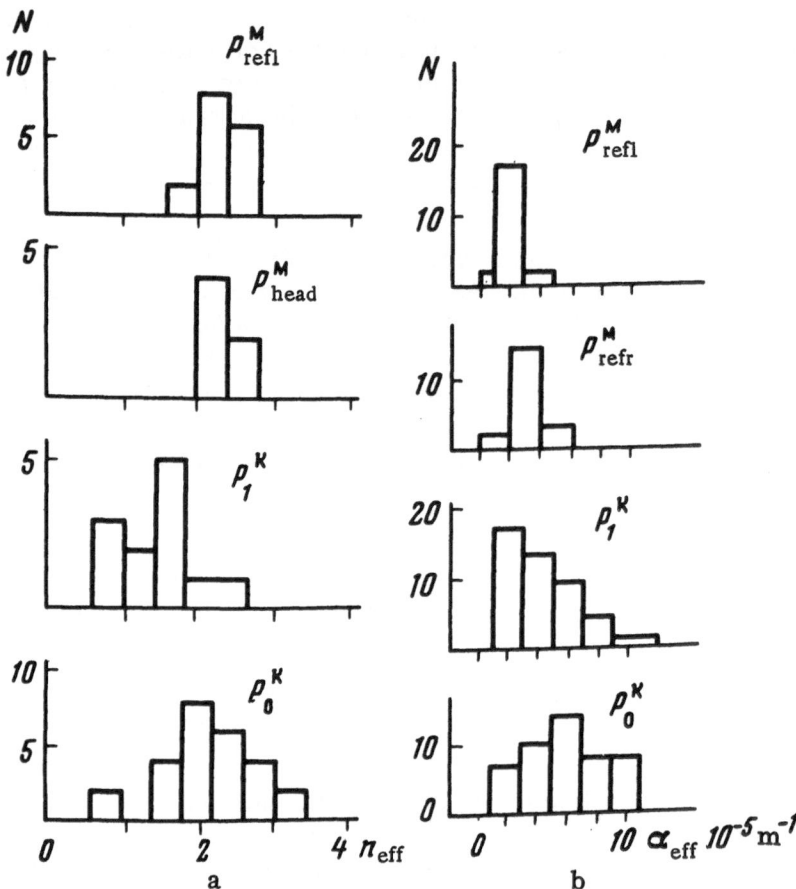

Fig. 4.14. Distribution of values for n_{eff} and α_{eff} for A(R) curves (Sea of Okhotsk [163, Chapter 4]): a) n_{eff} ; b) α_{eff} ; N) number of determinations.

Examples of histograms for values of n_{eff} and α_{eff} obtained from A(R) curves from the northern and central parts of the Sea of Okhotsk are shown in Fig. 4.14. These histograms were used in determining the nature of the crustal and mantle waves, K and M, in this area [163].

§4. Interference Zone in the Vicinity of the

Critical Point

Study of the phenomenon of interference is of fundamental interest in deep seismic sounding because interference between wave groups is observed over long intervals for the frequency ranges used on land (8-12 Hz) and at sea (4-6 Hz). Qualitative and quantitative analysis of the interference between two waves is not particularly difficult if the characteristics of the constituent waves are known. The inverse problem, that of determining the properties of the interfering waves (velocity and amplitude), is also solvable [93].

With a multiple layered model of the medium, it is possible to consider various types of interference between two waves. In so doing, the superposition of waves in the vicinity of the critical point is of considerable significance.

Interference between waves as a consequence of the intersection of travel-time curves may take place at distances passing through the critical point. In such cases, the amplitude characteristics of each of the waves may be expressed as an asymptotic approximation. Such interference is characteristic for the majority of deep waves in the range of first arrivals over the interval 50-200 km.

However, near the critical point, the asymptotic expressions for the amplitudes of interfering reflected and refracted waves are not valid.

A third case is also possible — the interferences between three waves, of which two (reflected and refracted from a single interface) are being traced in the vicinity of the critical point and represent, essentially, additive waves and the third is well away from its critical point and may be represented by an asymptotic expression.

We will consider the first two cases:

1. Interference of Two Waves Far from the Critical Point. The interference of two successive waves has already been considered in Chapter III, Section 3. The characteristics of such an effect for the case of the crust and upper mantle are a smooth and consistent relationship between the relative intensities and velocities of the constituent waves at various distances from the source.

The intensity $A_\Sigma(R)$ and apparent phase velocity $V_\Sigma(R)$ of the resultant oscillation for waves with a nearly sinusoidal form may be determined within a reasonable degree of accuracy using the formula for the case of the combination of two harmonic waves t_1 and t_2 [90]:

$$A_\Sigma(R) = A_2(R)\sqrt{1 + \eta^2(R) + 2\eta(R) \cdot \cos\omega\,[t_1(R) + t_2(R)]}\,, \qquad (4.2)$$

$$V_\Sigma^\bullet(R) = V_1^\bullet(R)\,\frac{1 + \eta^2(R) + 2\eta(R)\cos\omega\,[t_1(R) - t_2(R)]}{k(R) + \eta^2(R) + \eta(R)\,[1 + k(R)]\cos\omega\,[t_1(R) - t_2(R)] + \dfrac{\eta'\,V_1^\bullet(R)}{\omega}\sin\omega\,[t_1(R) - t_2(R)]}\,, \qquad (4.3)$$

where $A_1(R)$ and $A_2(R)$ are the amplitudes (these may be determined experimentally or computed from asymptotic expressions), $V_1^\bullet(R)$ and $V_2^\bullet(R)$ are the apparent velocities of the constituent waves, $k(R) = [V_1^\bullet(R)]/[V_2^\bullet(R)]$; $\eta = [A_1(R)]/[A_2(R)]$; $t_1(R)$ and $t_2(R)$ are the travel times, $\eta'(R)$ is a derivative with respect to R, and ω is the angular frequency. There may be a phase difference as well as a time difference between two interfering waves, related to a change in the phase characteristic for reflected waves in the vicinity of the critical angle.

It follows from an analysis of these expressions that the intensity of a complex wave varies periodically with distance and has extremals — maxima and minima where the amplitudes of the vibrations being combined add or cancel. The curve $A_\Sigma(R)$ oscillates about the amplitude curve for the dominant wave. The phase travel-time curves for the combined wave also oscillate about the travel-time curve for the dominant wave and in addition, the phase velocity depends on the ratios $\eta(R)$ and $k(R)$, and may be positive, negative, or infinite in the neighborhood of the maximum amplitudes.

For a constant ratio of amplitudes, $\eta(R)$ = const and $k(R)$ = const; that is, for straight-line travel-time curves, Eq. (4.3) simplifies to:

$$V_\Sigma^\bullet(R) = V_1^\bullet\,\frac{1 + \eta^2 + 2\eta\cos\omega R\,[1/V_1^\bullet - 1/V_2^\bullet]}{k + \eta^2 + \eta(1 + k)\cos\omega R\,[1/V_1^\bullet - 1/V_2^\bullet]}\,.$$

At extremal points on the amplitude curve, where $A_\Sigma(R) = A_1(R) \pm A_2(R)$, we have:

$$V_\Sigma^*(R) = V_1^* \frac{\eta \pm 1}{\eta \pm k}. \qquad (4.4)$$

With the negative sign in Eq. (4.4) and with $\eta = k$, the value for $V_\Sigma^*(R)$ becomes infinite; with $\eta = 1$, it is zero, and with $k > \eta > 1$ or $k < \eta < 1$, it is negative.

At points where $A_\Sigma(R) = A_1(R) + A_2(R)$, that is, in areas with local maxima in amplitudes (this corresponds to sections where records show a better degree of correlatability and a simpler wave form for the interfering waves), the velocity $V_\Sigma^*(R)$ is always positive and is

$$V_\Sigma^*(R) = V_1^* \frac{\eta + 1}{\eta + k}.$$

This expression has already been used in §3.

We shall now develop several simple relationships for the characteristics of complex waves which are of interest in the recognition of interference effects.

The distance ΔR between a maximum and a minimum in the amplitude curve with η and k = const is measured in wavelengths $\lambda = V_1 T$, where T is the period of the oscillation, and k is:

$$\Delta R = \frac{\lambda}{1 - k}.$$

The maximum departure Δt_Σ^* between the travel-time curve for the combined wave and the travel-time curve for the dominant wave may be expressed in terms of the period T for $k < 1$ and $\eta > 1$,

$$\Delta t_\Sigma = -T \frac{\eta}{1 + \eta}, \qquad (4.5)$$

and for $\eta < 1$ and $k < 1$,

$$\Delta t_\Sigma = T \frac{\eta}{1 + \eta}. \qquad (4.6)$$

It is evident from (4.5) and (4.6) that the departure of the travel-time curves does not depend on the velocity ratio, but is determined only by the ratio of amplitudes of the two constituent waves. The size of Δt_Σ is always less than a half period.

Curved travel-time curves for complex waves are approximated with straight-line segments, and if the dominant wave has the lower velocity, the travel-time curves have rising steps which correspond to the shift from one phase to another. If, on the other hand, the dominant wave has the higher velocity, the travel-time curve for the combined wave will have descending steps. In the first case, the velocity of the combined wave will be higher than the velocity of the dominant wave, and in the second case, lower.

The first type of travel-time curve is characteristic of a complex wave in the zone of interference between a low-angle reflection and a refracted wave, where the refracted wave is dominant, while the second type is characteristic of the region of overlap between early waves in groups.

In analyzing deep seismic sounding records, as well as in other branches of seismology, in most cases it is quite reasonable to recognize the diagnostic features of interference and delineate the boundaries of such zones, but quantitative evaluation of complex vibrations is made difficult by the superposition of various forms of the noise, and more importantly, by nonuniformity of the medium, as was discussed in §3.

The most useful indications are distortion of the nearly sinusoidal forms of the recorded waves and the form of the amplitude curve. In many areas, the amplitude curves for deep wave groups are complicated by alternating extremals [3, 155, 163] which are especially notable on low-frequency records and which confirm the presence of intergroup or intragroup interference.

2. Interference between Reflected Waves and Head Waves in the Vicinity of the Critical Point. This case is of primary interest because in most areas the reflected waves from deep boundaries (particularly boundaries in the lower part of the crust and the upper mantle) are most clearly recognized and traced most readily in the vicinity of the critical point.

Both theoretical [157, 158, 195-200] and experimental [69, 71, 73, 74, 89, 98] studies of the behavior of waves in the vicinity of the critical point have been described in the literature.

The difficulty involved in such studies is that from the physical point of view, beginning at distances corresponding to the abscissa of the critical point, there is only one wave formed which is gradually deformed, and only at some distance from the critical point can it be resolved into two different vibrations: a reflected wave, and a refracted head wave.

The characteristics of the combined wave in this region may not be expressed in terms of asymptotic formulas for each of the constituent waves, as is done at large distances from the critical point; near the critical point, the function $L = R^{-1/2}(R - R_j)^{-3/2}$ for the propagation of the head waves tends to infinity.

Červeny (Cherveni) [195, 199] and Smirnova [157, 158] have used various approaches for studying the characteristics of the complex waves, which may be considered as varying degrees of approximation to the exact solution of the wave equation in the vicinity of the critical point. They have shown that the amplitude characteristics of the combined waves may be expressed:

a. with the use of asymptotic expressions; that is, with a geometric approximation, if the reflected and refracted waves are added in this region with consideration of phase shift [195];

b. with the use of a more exact approximate expression for harmonic vibrations, in which the variation of the phase characteristic for the reflected and refracted waves and their interference are considered [195]; and

c. with the use of exact numerical solutions of the weak equation for an arbitrary form of the incident pulse (Smirnova [158]).

In determining the practicality of the solutions involving various degrees of approximation, we will compare the amplitude curves and synthetic seismograms computed with the various formulas.

Amplitude Curves. Figure 4.15 shows the results of combining the intensities of harmonic reflections and head waves computed from the asymptotic expressions with a phase shift of $\pi/2$ at the critical point being assumed and a time difference which enters with distance from the critical point. A diagnostic property for the A_Σ curve for the combined wave is the presence of extremals, and moreover, the principal extremal is the first maximum, which is shifted toward larger distances from the geometrical position of the critical point.

The curves for the reflected wave A_0 and the head wave A_H are shown on the same illustration. It should be stressed that the curve for head waves has been added to the curve for reflected waves at points shifted to the right from the critical point. Curves have been published elsewhere frequently in which the curves are added without shifting the critical

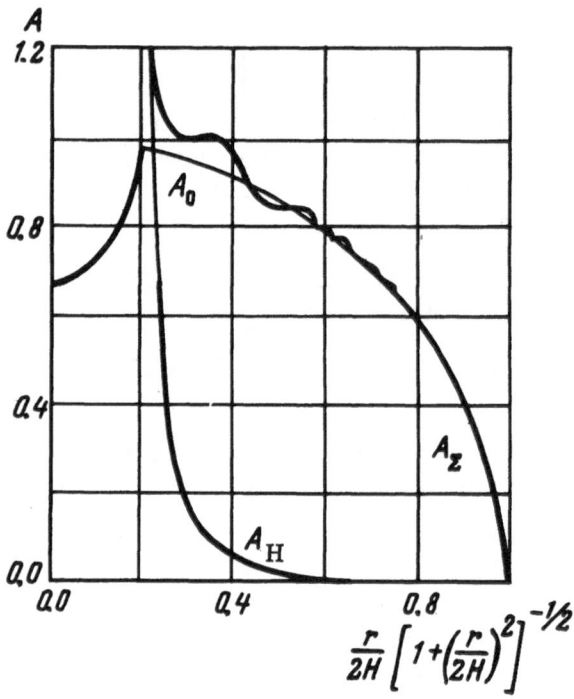

Fig. 4.15. Amplitude curves for the combination of
a reflected and head wave A_Σ in the vicinity of the
critical point. The curve A_Σ was constructed by
adding the reflected wave A_0 and the head wave A_H
with a phase shift. The source, harmonic behavior,
fluid medium, and effect of the earth's surface have
not been considered. The ratio of velocities is n =
0.2, and the ratio of densities is $\sigma = 1$; r is the distance from the source; H is the depth to the boundary
[195, p. 331, Fig. 13].

point; these are obviously incorrect inasmuch as the asymptotic expressions give an amplitude for the head waves which tends to infinity at the critical point.

With the next better degree of approximation, considering the change in the reflection
coefficient near the critical point and the exact expression for head waves [196], the curves differ from the preceding case by being smoother and at a somewhat lower level for the intensity of the reflected wave in the transition to the geometrically exact location of the critical
point (Fig. 4.16).

The latter extremals on the curves for the combined waves play a minor role. Because
of the decrease in intensity of the head waves, even at short distances from the critical point,
the average curve is the curve for the dominant wave A_0. The curves shown in Fig. 4.15 were
computed for a liquid medium. The form of the curves is not much different for the case of
a solid medium [198].

It is obvious that the principal extremal is of the most interest in evaluating the reflecting
and refracting properties of an interface. An analysis of its location and form as functions of
the properties of the medium and recorded frequencies is illustrated by the following graphs.

Figure 4.16 shows graphs of the intensity A_Σ for the various ratios n for the velocities
of longitudinal waves at the boundary. The larger n is, the broader and flatter will be the
maximum.

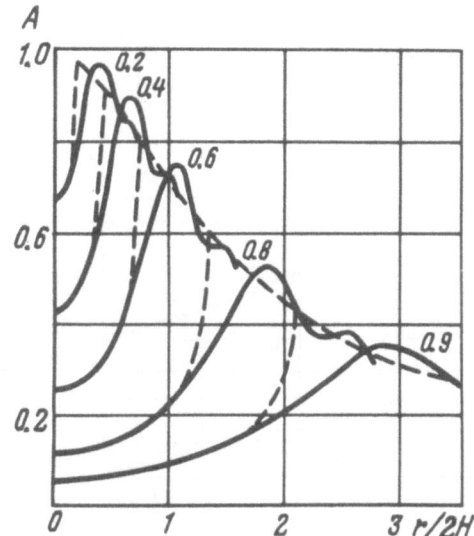

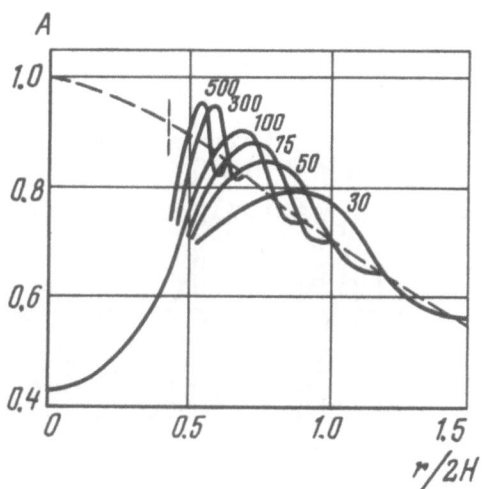

Fig. 4.16. Relationship of the amplitude curves for interfering reflected and head waves to the velocity ratio n. The dashed lines are the amplitudes for reflected waves computed from asymptotic expressions; the parameters are $\sigma = 1$; $4\pi H/\lambda = 100$, λ is the wavelength; the rest are the same as in Fig. 4.15; the figure with each curve is the value for n [195, p. 340, Fig. 18].

Fig. 4.17. Amplitude curves for the combined wave A_Σ with curve parameters being H/λ; the rest of the parameters are the same as in Fig. 4.15 [195, p. 341, Fig. 21].

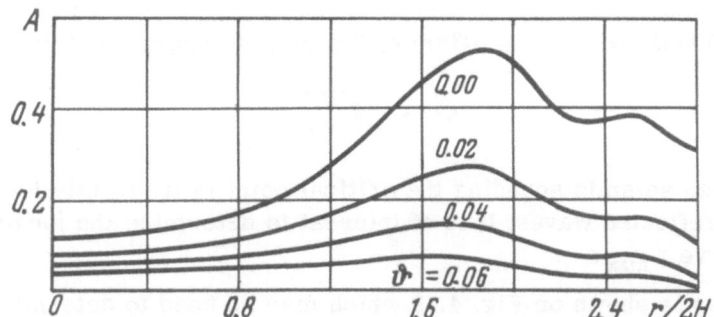

Fig. 4.18. Relationship of the form of the amplitude curve for the interfering waves to attenuation in the upper medium [195, p. 345, Fig. 24]. The curve parameter is the attenuation decrement $\vartheta = \alpha\lambda$, where α is the attenuation factor and λ is the wavelength; $n = 0.8$, $\sigma = 1$, and $4\pi H/\lambda = 100$.

The form of the maximum is relatively independent of the density contrast. However, it does change significantly for various ratios H/λ (H is the depth to the interface and λ is the wavelength in the upper layer, Fig. 4.17). The larger H/λ is, i.e., the higher the frequency is, the narrower the maximum becomes and the closer it is found to the geometrically exact critical point.

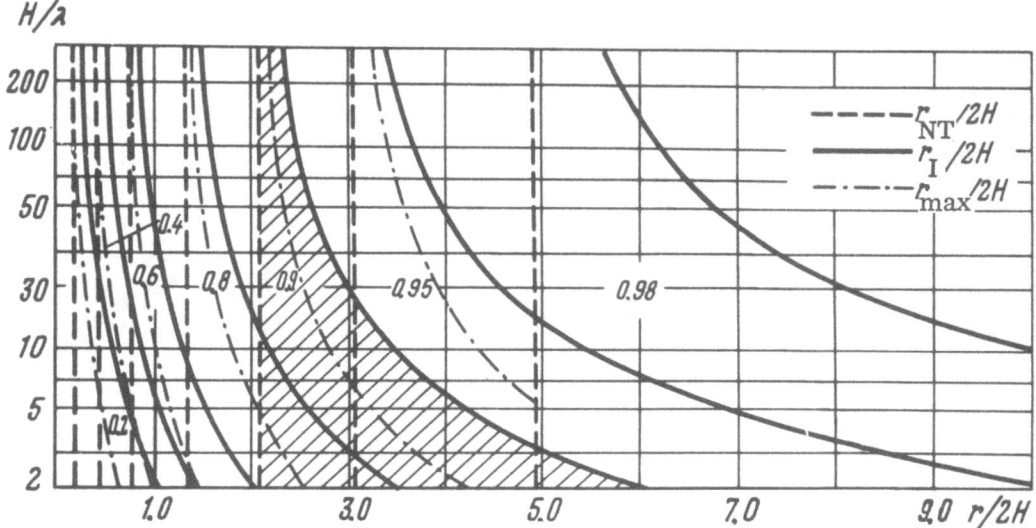

Fig. 4.19. Curve set for determining the extent of the interference zone.

Consideration of attenuation in the upper medium smooths out the intensity curve for the interfering waves. However, for the small values of the attenuation factor in the earth's crust, where it apparently is no more than 0.01, the effect of attenuation on the form of the interference maximum need not be considered (Fig. 4.18).

Extent of the Interference Zone. In evaluating the extent Δr_I of the interference zone where it is not permissible to examine the reflected and head waves separately, we may use the expression [196, p. 56]:

$$\Delta r_I = \frac{\lambda}{1-n^2}\left(n + \sqrt{4H/\lambda \sqrt{1-n^2}+1}\right).$$

For $H/\lambda > 10$ and $n < 0.9$, the extent of the zone is approximately:

$$\Delta r_I \approx \frac{2\sqrt{H\lambda}}{(1-n^2)^{3/4}}.$$

Because in deep seismic sounding the critical point is frequently identified by the appearance of strong reflected waves, it is of interest to determine the location of the maximum on the combined curve r_{max}.

A set of curves is shown on Fig. 4.19 which may be used to determine r_{NT}, r_I — the abscissa of the end of the interference zone, and r_{max} when n and H/λ are given. In the shaded area, $n = 0.9$. For example, with $H/\lambda = 30$, $r_{max} = 2.4 \times 2H = 4.8H$, $\Delta r_I = r_I - r_{NT} = (2.6 - 2.1) \times 2H = H$.

The curve r_{max} (H/λ) is situated essentially in the middle of the interference zone. Therefore, it may be assumed that the amount of displacement of the maximum on the A_Σ curve with respect to the geometrically exact position of the critical point is approximately $\Delta r_I/2$; that is, half of the width of the interference zone. The shift may be computed more exactly with the expression [198, p. 287]

$$r_{max} - r_{NT} = \beta\sqrt{\lambda r_{NT}},$$
$$\beta = 0.86n^{-1/2}(\lambda - n^2)^{-1/2}.$$

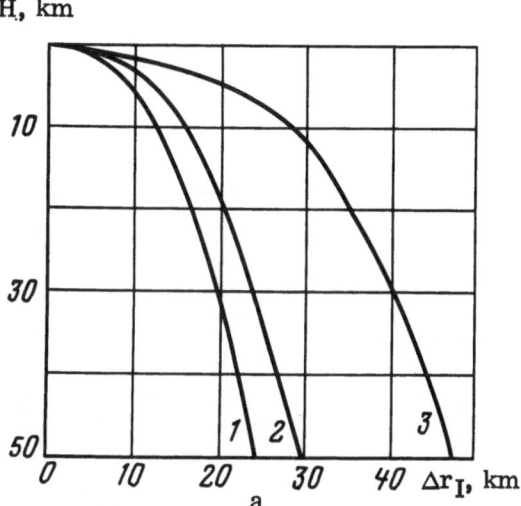

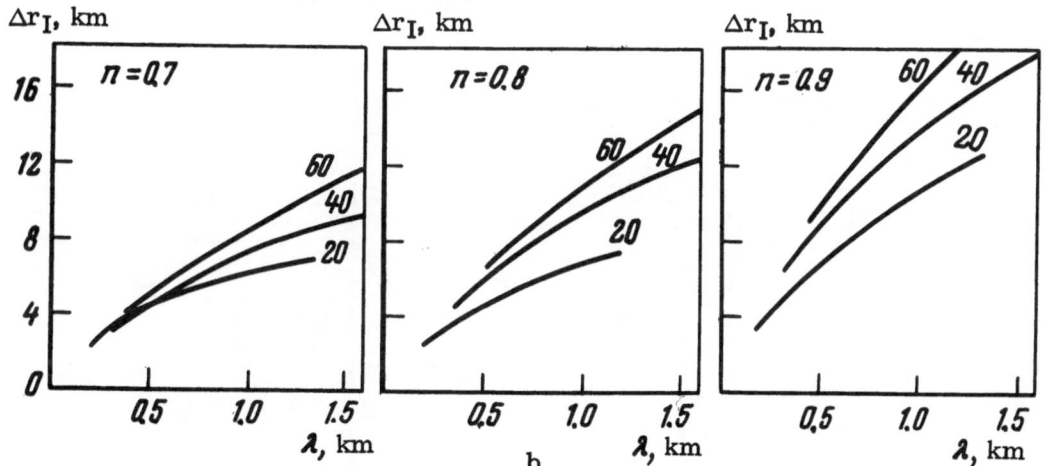

Fig. 4.20. Width of the interference zone Δr_I. a) As a function of the depth H^M to the M-discontinuity for various types of crust: 1) n = 0.75, V_1 = 6.0 km/sec; 2) n = 0.825, V_1 = 6.6 km/sec; 3) n = 0.94, V_1 = 7.5 km/sec, V_2 = 8 km/sec; b) as a function of wavelength; the curve parameters is $2H/\lambda$.

Values of β for various n are given below:

n	0.30	0.40	0.50	0.60	0.70	0.80	0.90	0.95
β	1.65	1.51	1.40	1.39	1.44	1.60	2.08	2.83

Figure 4.20 shows graphs for the relationship between the width of the interference zone Δr_I and the depth H to the M-discontinuity and the velocity ratio n. The curves have been computed for a series of multiple layered crustal sections. In the calculations, it was assumed the $\lambda = \Delta t V_{k-1}$, where V_{k-1} is the velocity in the layer on top of the M-discontinuity and the wavelength is $\Delta t = 0.25$ sec.

It is evident from these illustrations that the graphs for the intensity of the combined waves differ significantly from the graphs for reflected and head waves. It is close in form to the graph for the reflected wave, which is dominant in amplitude, but the extremal characterizing the reflected wave in the vicinity of the critical point is shifted with respect to the ab-

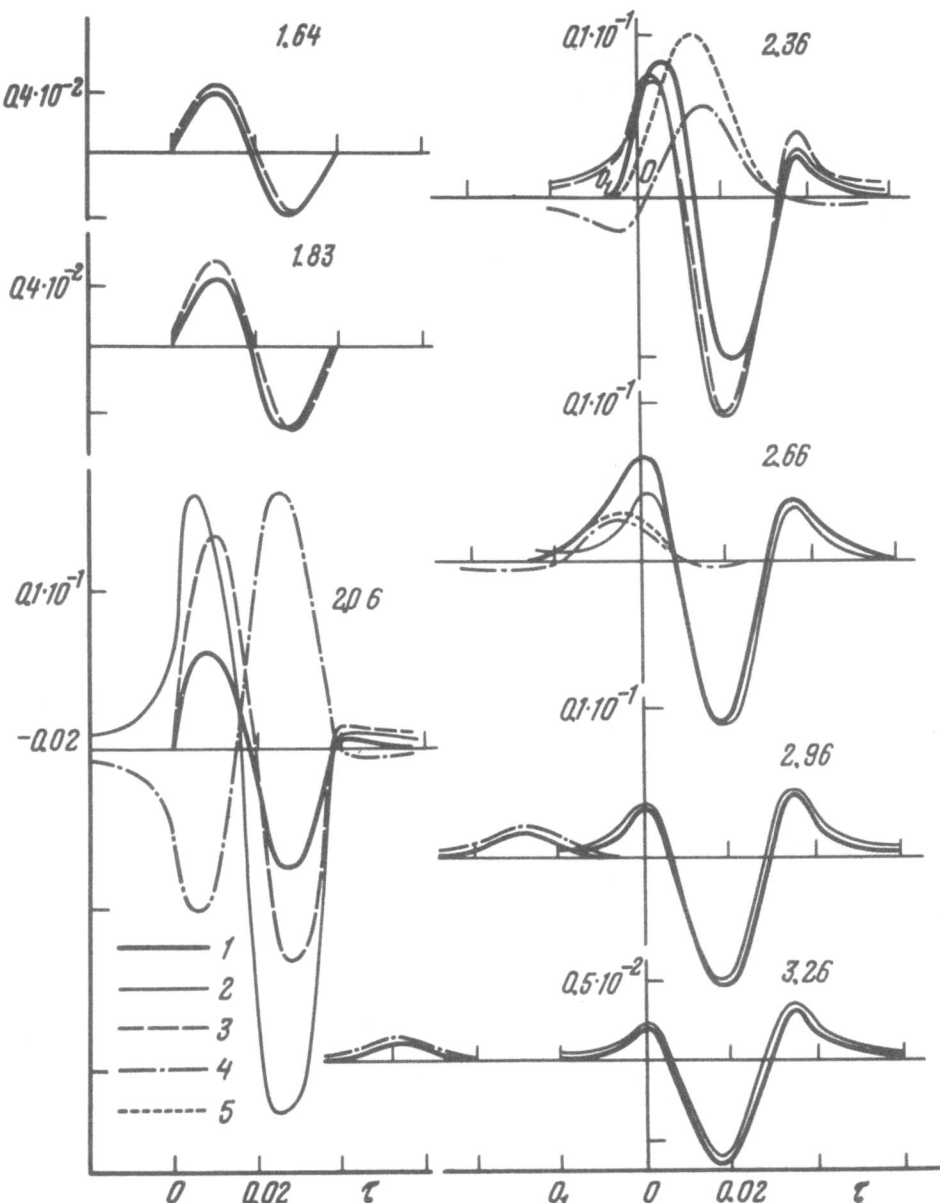

Fig. 4.21. Synthetic seismograms computed by N. S. Smirnova from exact [158] and asymptotic [43] formulas (vertical component): 1) combined wave; 2) A_0 reflected wave from the exact formula; 3) the same, from the asymptotic formula; 4) A_H head wave from the exact formula; 5) the same, from the asymptotic formula. The curve parameter is $r/2H$, the velocity ratio is $n = 0.9$, t is time in sec, r is distance, H is the depth to the interface, $r/2H = 2.06$ corresponds to the neighborhood of the critical point.

scissa of the critical point as a result of the effect of the head waves. With greater distance from the critical point, the combined vibration grades into a reflected wave.

Despite the fact that the representation of the reflected and head waves in the interference zone is only possible as a formal concept, evaluation of the behavior of each of these waves is of practical interest in analyzing the feasibility of recording the first vibrations of the head waves as they emerge from the interference zone.

TABLE 17. Ratios of Amplitudes of Reflected and Head Waves
at the Point r_{max} and at the End of the Interference Zone r_I

n \ H/λ	4	9	25	49	100
0.7	0.42	0.34	0.27	0.22	0.19
	0.17	0.14	0.11	0.09	0.08
0.8	0.39	0.32	0.25	0.21	0.18
	0.16	0.13	0.10	0.08	0.18
0.9	0.36	0.30	0.23	0.19	0.16
	0.15	0.12	0.09	0.08	0.07

Note: The upper figures pertain to r_{max} and lower to r_I.

It has been found from a comparison of asymptotic and exact solutions (Fig. 4.21) that the asymptotic expressions may be used for calculating the ratio of amplitudes of reflected and head waves beginning at distances corresponding to the abscissa of the first maximum of the combined waves (see Fig. 4.15). Thus

$$A_I/A_0 = 0.18 F(n) \cdot (H/\lambda)^{-1/4},$$

where $F(n) = n^{-1/2} \cdot \Gamma^{PP}(P)(1 - n^2)^{1/8}$, and the function $\Gamma^{PP}(P)$ has been tabulated in [43].

Table 17 is a list of the computed amplitude ratios for various H/λ and n which are of interest in deep seismic sounding.[†] The upper figures correspond to the location of the maximum on the A_Σ curve, and the lower figures correspond to the end of the interference zone. The calculations were done for a sinusoidal pulse with a length of 0.25 sec.

The ratio A_H/A_0 depends primarily on H/λ, that is, on frequency. With a tenfold increase in frequency, the relative intensity of the head waves decreases by a factor of almost two. A minor decrease in the ratio A_H/A_0 with increasing n is caused by a marked broadening of the maximum and a shift away from the initial point. Thus, the head wave, because of its rapid rate of decay is suppressed more strongly than the less rapidly decaying reflected waves, despite the fact that the relative intensity of the head waves near the initial point increases with increasing n.

It is of interest to compare the ratio A_H/A_0 at the maximum point and at the end of the interference zone between reflected and head waves. The ratio is less than 0.1 at the end of the zone, or less than the value at the maximum by a factor of about 2.

Wave Forms and Wave Spectra. Computed seismograms for Gaussian (Fig. 4.21) and sinusoidal (Fig. 4.22) pulses based on the exact formulas indicate that some changes in the recorded wave forms take place in the vicinity of the critical point with interference between reflected and head waves: increase of pulse width, increase in the relative intensity of the later phases, and increase in periods. However, over some frequencies and some ranges for the properties of the medium, these changes are not so large as to inhibit correlation of the primary extremals within the interference zone. The wave form of the record made right at the critical point is much the same as the form of the reflected wave at higher and lower angles (but near the critical point).

An examination of the amplitude spectra for the waves (Fig. 4.22) in the vicinity of the critical point indicates how insignificant the changes at the critical point are.

[†] Table 17 was prepared by V. Červeny (Cherveni) under the direction of the author.

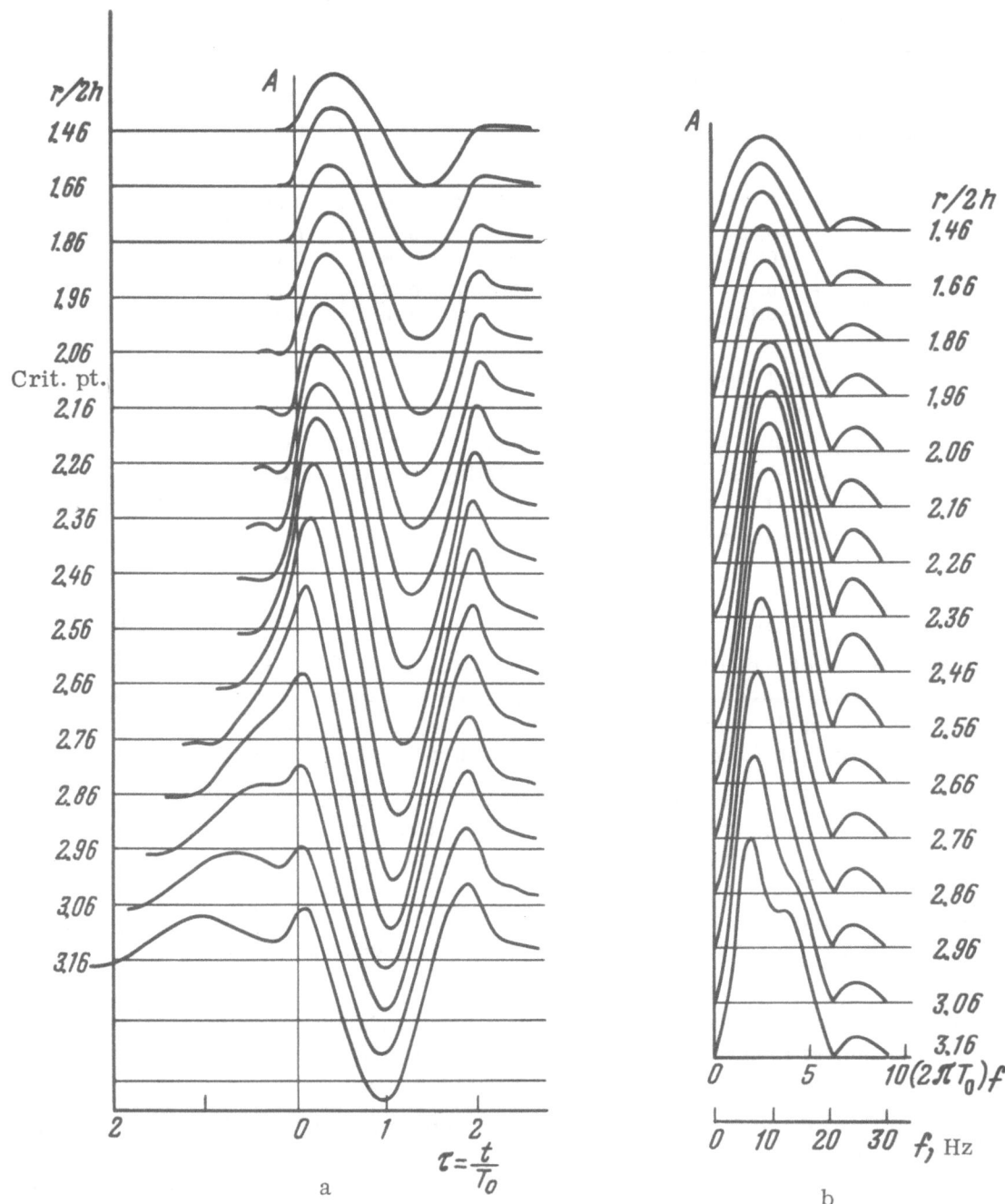

Fig. 4.22. Seismograms (a) and spectra (b) of a complex wave in the vicinity of the critical point, from calculations by N. S. Smirnova. T_0 is a half period of the incident wave; the low scale for the spectra (f, Hz) is given for $T_0 = 0.05$ sec; computations were done for n = 0.9, $\sigma = 1$, $2H/\lambda = 30$, $\gamma = V_s/V_p = 0.6$.

Deep seismic sounding records showing clearly the behavior of waves near the critical point are shown in Fig. 4.23.

Phase Travel Times and Velocities. We will analyze the form of the travel-time curve for the maximum phase in a complex vibration. It may be shown using raypath theory that phase

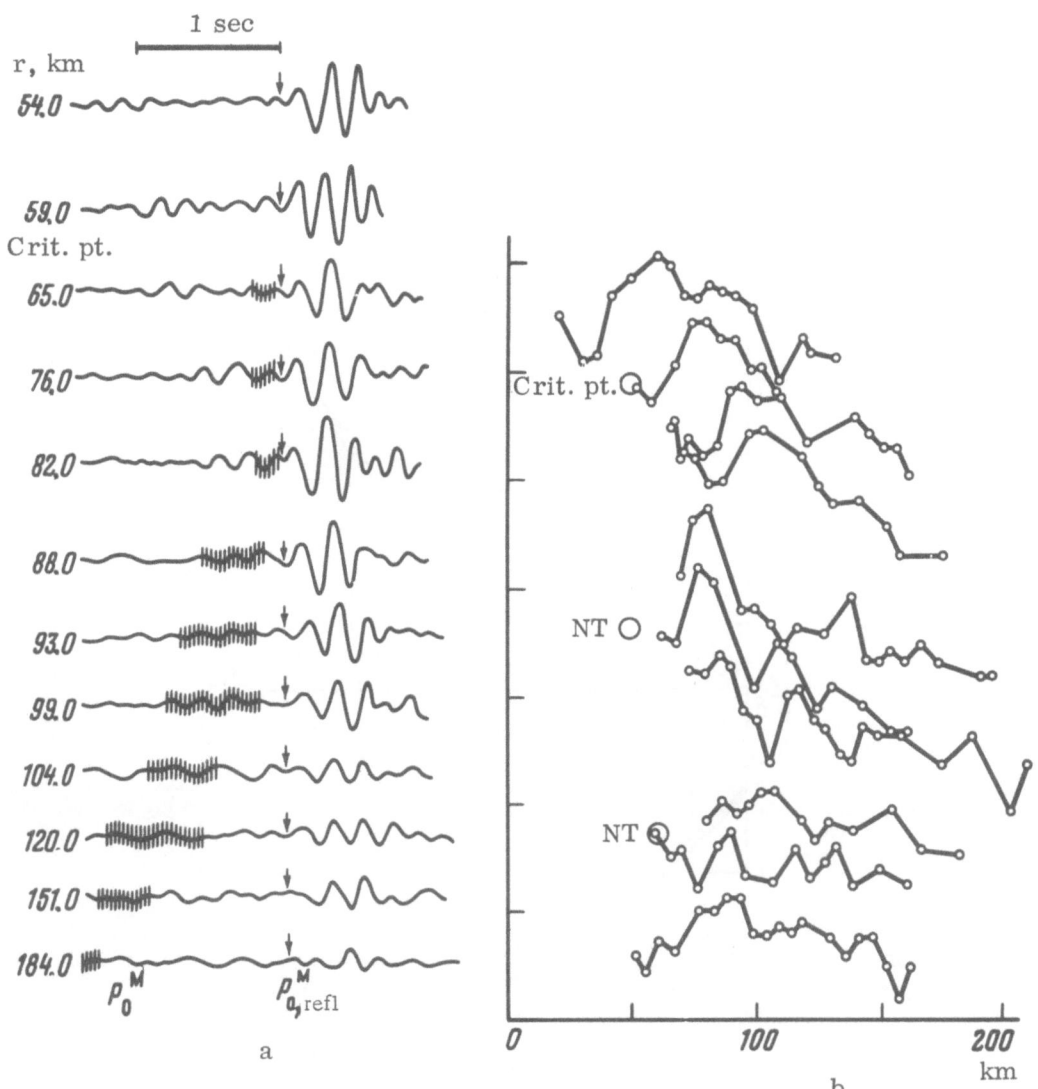

Fig. 4.23. Seismograms (a) and amplitude graphs (b) in the vicinity of the critical angle for the $P^M_{0\,refl}$ wave (Sea of Okhotsk [163, Chapter 7]). Crit. pt. – location of the critical point; the shaded areas on the seismograms indicate the position for head waves.

shift is present for the reflected wave only from the critical point out, and rarely varies with respect to rate of change (Fig. 4.24a, curve 1). Exact theory gives somewhat different results. According to calculations, phase shift of reflected waves takes place even ahead of the critical point, smoothly increasing with distance. The curve for the relationship between phase shift and distance (Fig. 4.24a, curve 2) passes through the critical point without any change in curvature. At greater distances, the phase shift oscillates about the corresponding curve for reflected waves. Phase travel-time curves for interfering vibrations may be constructed on the basis of these calculated phase shifts [90, 185].

Travel times for the first arrival and phases in passing through the critical point, computed both from exact formulas and the raypath approximation, are shown in Fig. 4.24b.

The phase travel-time curves are not parallel to the travel-time curve for the beginning of the wave packet, and in addition, this nonparallelism is greater for the exact calculations.

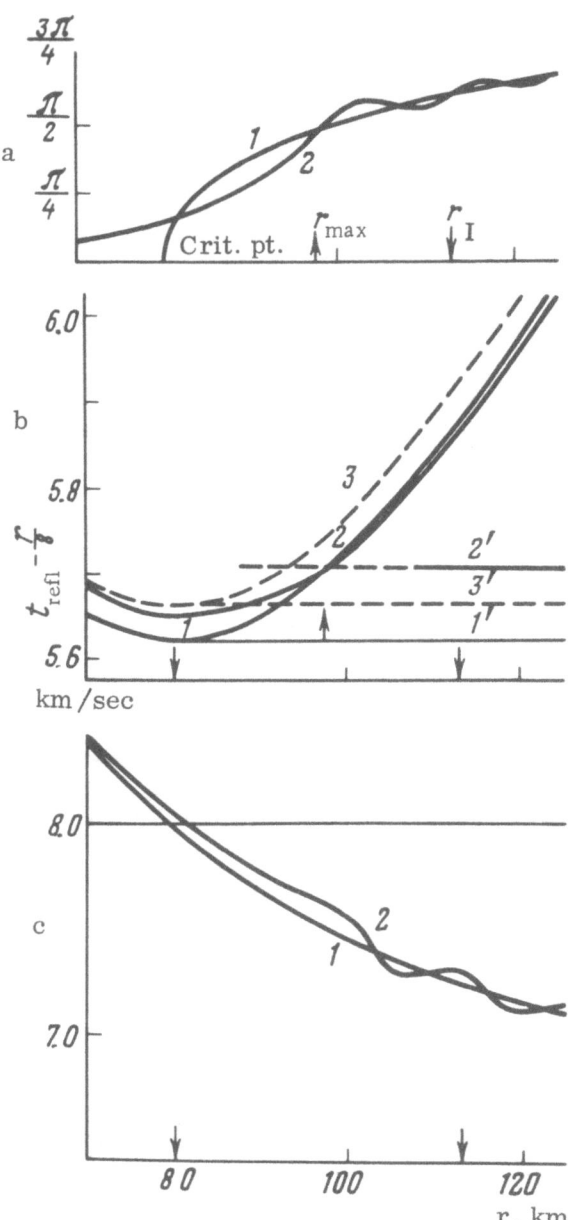

Fig. 4.24. Phase characteristics for complex vibrations [185]. a: 1) Phase shift for reflected waves; 2) for combined waves. b: Phase travel times: 1–1') travel times for first arrivals; 2–2') travel times for the maximum phase of a combined wave; 3–3') phase travel-times for reflected and head waves. c: Apparent velocities: 1) reflected waves; 2) phases of a complex vibration. Crit. pt.) critical point; r_{max}) abscissa for the maximum amplitude; r_I) end of the interference zone.

A characteristic feature of phase travel-time curves in an interference zone is that the travel-time curves for the reflected and head waves are not distorted over their lengths (as is the case for the travel-time curve for the beginning of a wave packet), but intersect. The intersection point is located at almost the same position as that for the maximum amplitude for the combined vibrations. This is important in estimating the accuracy of locating the critical point from phase travel-time curves as compared to the travel-time curves for the beginning of the wave packet.

Curves showing the change in apparent velocity with distance for the beginning of a wave packet and for the maximum phase for low-angle and intefering high-angle waves are shown in Fig. 4.24c (for the travel times in Fig. 4.24b).

The apparent velocity of the maximum phase in a reflected wave computed from exact formulas very close to the critical point is larger than the velocity for the beginning of the wave packet.

For higher-than-critical angles, the curve for the apparent velocity of the maximum phase oscillates about the curve for the beginning of the wave packet; that is, in this range, the apparent velocity of higher-than-critical-angle reflected waves may be somewhat larger or somewhat smaller than the velocity of the reflected wave. This departure is larger at lower frequencies. This difference does not exceed ±0.1 km/sec in most cases in deep seismic sounding.

Beyond the critical angle, characteristically, the early phases of a reflected wave packet decrease more rapidly with distance than the later phases, and so in phase correlation, it is quite possible to slip to later phases of the vibration (see Fig. 4.24b).

In conclusion, we will make some comments. The range where asymptotic formulas, strictly speaking, are invalid begins before the critical point and ends when the transit times for reflected and head waves differ by an amount equal to the length of the head wave.

The distance to the left of the critical point amounts to about 0.3-0.4 of the length of the interference zone. The width of this zone is shown on Fig. 4.20b for the M-discontinuity as a function of the depth H, the velocity ratio n, and H/λ for various λ.

In practice, the asymptotic formulas may be used to calculate each of the waves, beginning at distances larger than the abscissa of the primary maximum of the combined waves, that is, essentially in the second half of the interference zone.

The recorded wave forms in the interference zone are not altered so much that the correlation of the primary extremals of the reflected waves is distorted. Initially, the combined waves propagate with the velocity of a head wave, but the phase velocity of the later vibrations is close to the velocity of propagation of the reflected wave.

As we have seen earlier, the intensity of the combined waves oscillates about the curve for reflected waves. In this respect, Červeny (Cherveni) [195] has suggested that the combined wave be considered to be a generalized reflected wave in the interference zone. I do not think that this term is entirely satisfactory, inasmuch as a fundamental concept is lost, that of the existence of head waves and their effect on the characteristics of the combined vibrations. It is obvious that in the vicinity of the critical point, it is best to preserve the concept of the combination of interfering waves which have properties that are determined primarily by the properties of the constituent waves, as in other cases in which vibrations are superposed. For the case of a uniformly layered model, such components are reflected waves for any value of n, while for a nonuniformly layered model, in the vicinity of the critical point, waves with similar amplitudes are involved, and either the reflected wave or the refracted wave may be dominant.

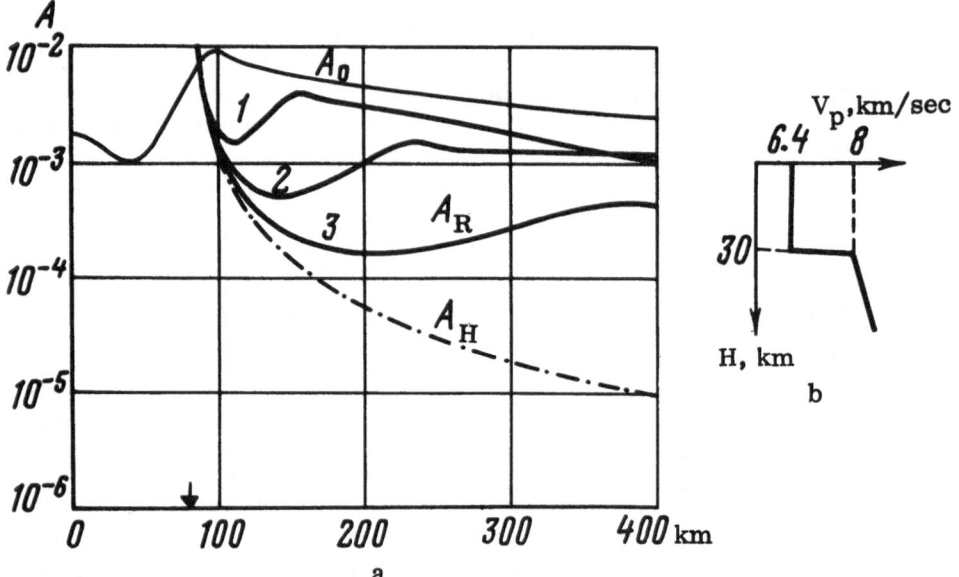

Fig. 4.25. Amplitudes for reflected (A_0), head (A_H), and geometrically reflected (A_R) waves in the vicinity of the critical angle [200]: a) curves 1-3 were computed from an exact formula by B. S. Chekin [183] for velocities increasing linearly with depth; the velocity gradients are 0.064, 0.023, and 0.008 km/sec/km, respectively. The arrow indicates the initial point for head waves; b) schematic velocity profile for which the amplitudes were computed; for H > 30 km, the velocity increases as: $V_L = V_0[1 + \beta(H - 30)]$; $f = 6.4$ Hz.

In deep seismic sounding, the nature of the interference in the vicinity of the critical point may differ for waves propagating in the upper part of the crystalline crust and in the lower part of the crust and upper mantle. In the upper part of the crust, the velocity gradients are larger and the intensity of geometrically refracted waves over the entire extent of the interference zone may be comparable with the intensity of the reflected waves. Equations (4.1) and (4.2) may be used to represent the amplitude and velocity of the combined wave approximately; that is, the phase velocity for $1 < \eta < 10$ will differ from the velocity for either wave.

In the lower crust and upper mantle the velocity gradients are lower, and the amplitude curve for the geometrically refracted wave will be close to the amplitude curve for a true head wave (Fig. 4.25). In this case, the general picture will be close to that considered earlier.

Practicality of Using Seismic Records Obtained in the Vicinity of the Critical Point. The region near the critical point is advantageous in terms of energy ratios for recognizing reflected waves against a background of later oscillations [42, 129, 155, 163, 224, and others]. The development of a theory for computing the amplitude and phase characteristics of seismic waves in passing through the critical point makes it possible not to exclude this zone from consideration, as is commonly done in studying wave forms, but to use data from this region for quantitative calculations, improving the techniques developed earlier solely on the basis of travel times.

This improvement is related primarily to the behavior of the critical point. Because reflected and refracted waves are usually recorded as later arrivals in wave packets at more-than-critical distances, phase correlation must be used in tracing them rather than the beginning times for the wave packets. For the seismic records and other curves which have been presented, we are not able to determine the nature of the critical point either on the basis

TABLE 18. Properties of Interference Zones

Type of zone	Relationship between r_{int} and r_{max}	Phase difference $(t_1 - t_2)$ at the point r_{max}
Head and head	$r_{int} = r_{max}$	0
Reflected + reflected	$r_{int} = r_{max}$	≈ 0
Head + reflected	$r_{int} > r_{max}$	$< \pi/2$
Reflected + head	$r_{int} < r_{max}$	$< \pi/2$
Critical point	$r_{crit} < r_{max}$	$\approx \pi/2$

Note: Initial phase shift at r_{crit} is $\pi/2$.

of wave forms, or from the nature of the amplitude curve or from the tangent point between the travel-time curves for reflected and refracted waves. These are not diagnostic characteristics at the critical point. However, at some specified distance from the critical point, the amplitude curve for a complex vibration has a sharp maximum.

As has been indicated by numerous experiments in various areas, the maximum of the amplitude curve is usually well expressed on records and its position may be determined with good accuracy [71, 98, 155, et al.]. The position of this maximum and its form depend on the properties of the medium. Knowing how this relationship must vary, it is possible to determine the range for recording reflected and refracted waves more satisfactorily.

Having the abscissa of the critical point and the transit time for the maximum phase of a complex vibration, any one of the properties of the medium may be evaluated quantitatively if the other properties have already been determined from the travel-time curves for reflected or refracted waves. For example, with values for the depth H, the velocity V, and the frequency f being known, it is possible to determine the velocity V_1 in the first layer, and so on.

Assuming not just one amplitude curve is available, but that a series of reversed curves or curves obtained at successive points along a profile in the survey area, it is possible to evaluate changes in the observed properties along the profile using changes in the position and form of the maximum.

The use of amplitude spectra of waves, which permits recognizing the relationship between curve parameters and frequency, is of considerable value.

Recognition of the Critical Point Region among Other Interference Zones. In conclusion, we will examine the diagnostic characteristics which may be used to recognize the critical point region among the other interference zones. Table 18 lists the various types of interference between two waves. All of the zones may be characterized by the relationship between the abscissa of the intersection point for the travel-time curves r_{int} and the point r_{max} where the combined wave has the maximum amplitude.

In cases in which the position of the maximum r_{max} coincides with the intersection point for the travel-time curves, that is, if the phase shift at this point is 0 or nT, where T is the period of the vibrations and n = 0, 1, or 2, then r_{max} will not depend upon frequency. This is a diagnostic characteristic for the interference zone for uniform waves from different boundaries (Fig. 4.26).

In case of interference between waves of different forms, the position of r_{max} may change with frequency. At lower frequencies, the point r_{max} will shift toward larger distances for all zones except those for which $r_{int} < r_{max}$. Computed curves for r_{max} as a function of fre-

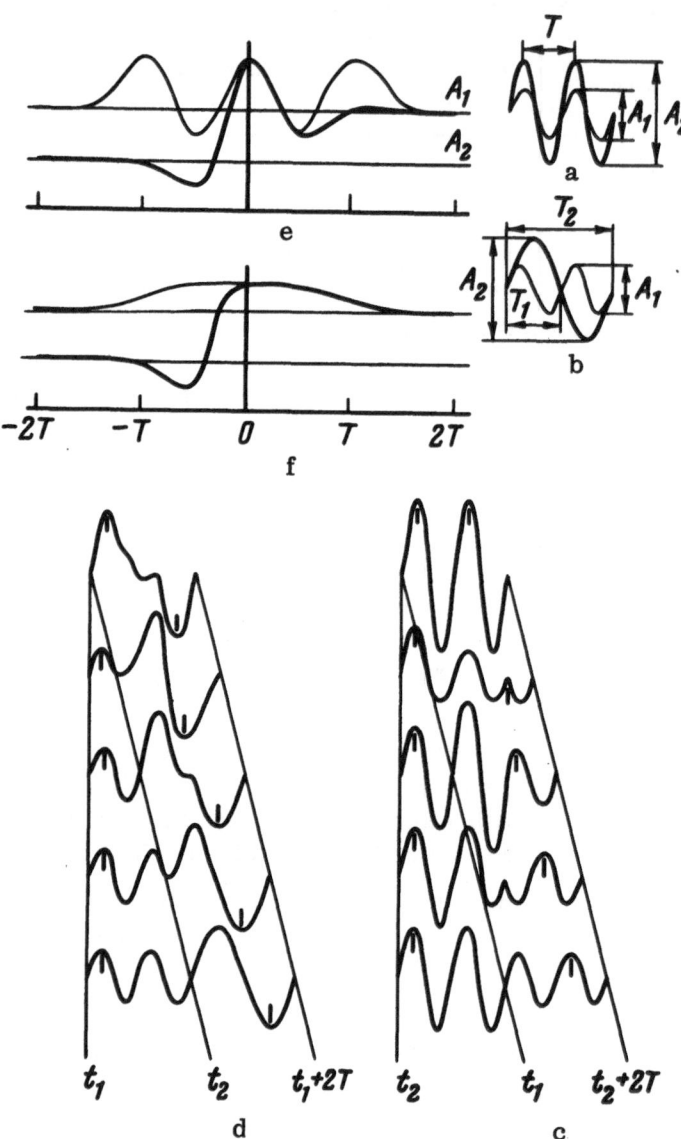

Fig. 4.26. Interference between two refracted waves
with different amplitudes and frequencies: a) form of
constituent oscillations of the same frequency; b) the
same, different frequencies; c and d) corresponding
seismic records for the combined waves; e) amplitude
graph for the constituent waves at a single frequency;
f) the same, different frequencies. The light lines correspond to the first phase of the complex vibration
while the heavy lines correspond to the maximum phase.

quency for several crustal models, presented by Shao Hsüeh-chung,[†] indicate that in the vicinity
of the critical point this shift is nearly twice as large as in an interference zone between nonuniform reflections and head waves corresponding to two adjacent boundaries. Shao Hsüeh-chung

[†] Shao Hsüeh-chung, Properties of Complex Waves in the Vicinity of the Initial Point in Deep
   Seismic Sounding, Cand. Dis., Inst. Fiz. Zemli, Akad. Nauk SSSR (1966).

also computed curves for the ratio of the spectra of one of the waves outside the interference zone to the spectrum of the complex waves. These curves are cosinusoidal in form.

Thus, in order to recognize the interference zone about the critical point positively among all other such zones, it is important to have not only the velocity characteristics of the complex waves, which we have discussed earlier, but also data on the relationship of wave form to frequency.

§ 5. Models of the Crust

The overall nature of the observed seismic waves in various tectonic zones which are relatively uniform in character corresponds in general terms to that expected for a horizontally layered medium, and the analysis of wave groups in zones where they can be traced reliably provides evidence that it is reasonable to divide this medium into a sequence of layers. However, in going from one region to another, the nature of the layering and the positions of the layer boundaries may change significantly, which is indicative of the blocky nature of the crust.

Within a gross block, the velocity profile is usually quite consistent, as indicated by the general characteristics of the seismic waves: the number of wave groups, and the ranges over which they are recorded. However, the dynamic characteristics of groups (relative intensities) and their internal structure (the character of individual waves) may change significantly along a traverse, and in particular, areas may be found where wave groups do not form. A comparison of these zones with geology (in Kazakhstan, the Baltic shield and the Ukranian shield) provides evidence that such zones are related to phenomena involved in development of the crust.

Studies of the features of seismic waves which characterize the surface of the crystalline basement, as well as zones where intrusive masses are found (Korosten pluton [159]; Pecheng structure [115]) now permit us to consider the question of the nature of the entire sequence of seismic waves, which we subdivide into wave groups in order to systematize the picture.

In discussing this question, we must incorporate geological considerations into our presentation (we will return to this later, also, in Chapter VI), and we will base this on the seismic data which have been presented above and from which it may be concluded that to a first approximation the average characteristics of each wave — the arrival time t_i, the apparent velocity, and the intensity at each point along a profile R_i — correspond to the average characteristics of seismic waves $A(R_i, t_i)$ for a multiple-layered medium in which the velocity increases with depth. The characteristics of wave groups correspond to the average characteristics of boundaries which separate uniform or weakly nonuniform velocity layers. Several models of uniformly and nonuniformly layered crustal models which have been obtained in various areas of the USSR are shown in Fig. 4.27, and amplitude curves for reflected and refracted waves were computed for these models using the raypath approximation.

Disagreements between the observed seismic waves and the properties of the models constructed from travel-time curves for wave groups consist of discontinuities in behavior, and primarily discontinuous correlation of the first waves in the packets.

Possible Causes of the Discontinuous Character of Seismic Waves. In seismic prospecting and in deep seismic sounding, a great deal of effort has been expended on the study of the nature of discontinuous correlations related to various causes [23, 24, 48]. In studies of the crystalline crust, data from the study of wave characteristics related to its upper boundary — the surface of the basement, which may have different structures in different areas, may be a complex of metamorphic and crystalline rocks, or intrusive masses of different types — are of particular

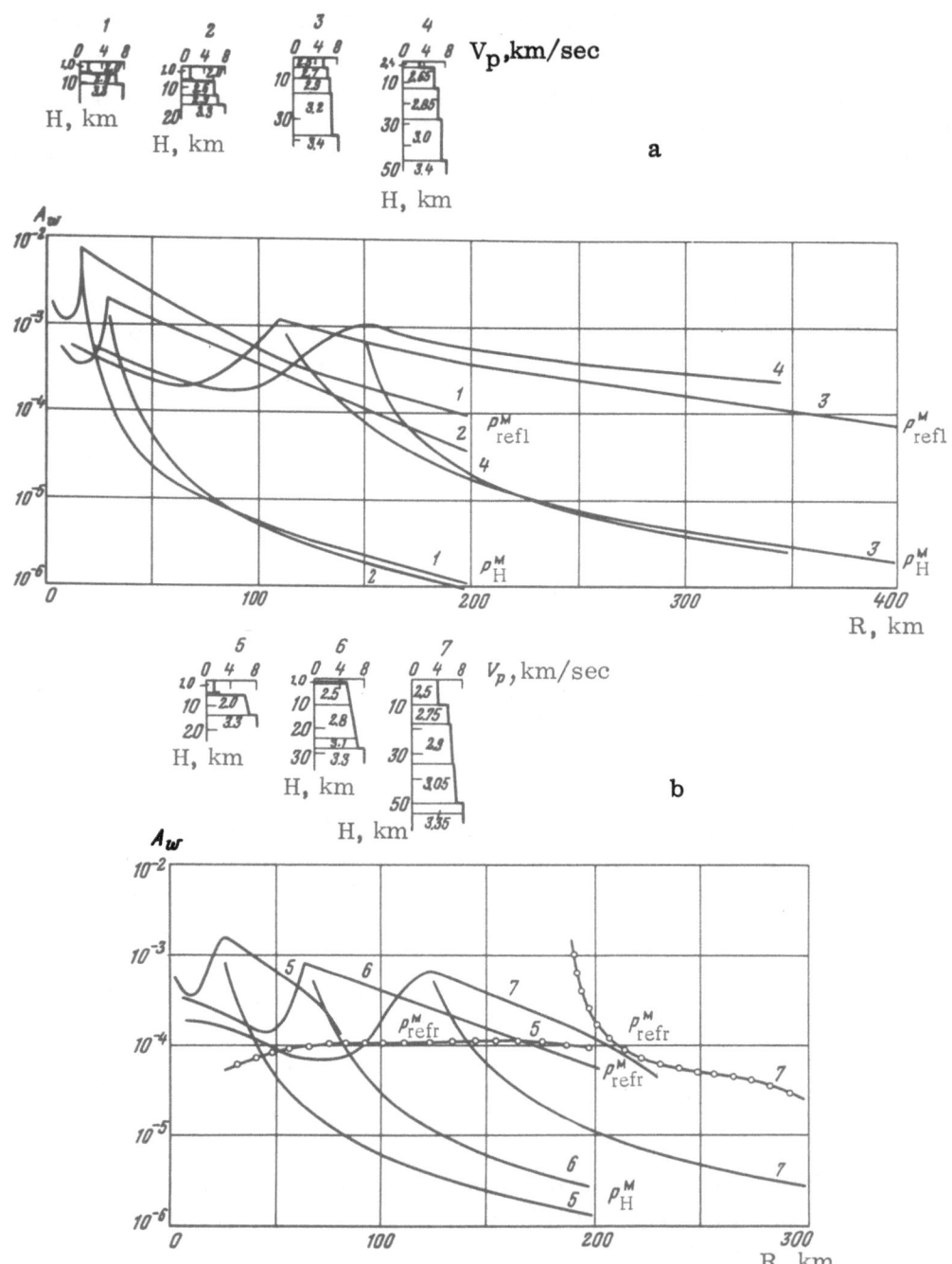

Fig. 4.27. Crustal models and amplitude curves: a) uniformly layered models; b) nonuniformly layered models. The figures in the sections are densities. The amplitudes along the ordinates are given in conditional units.

interest. Experiments in seismic prospecting with the refraction method in areas of basement outcrop and in covered areas have shown that rather than there being a single refracted wave associated with the basement surface, there is a group of waves which includes waves traveling along the surface and several waves which penetrate to small depths below the surface (weakly geometrically refracted), as well as diffracted waves, reflected waves, and reflected-refracted waves which are formed at the contact between two rock layers of different compositions [23,

24, 115, 116, etc.]. These waves interfere with each other, so that the first arrival in a group shifts, that is, there is a discontinuity in correlation. With continuous boundaries, discontinuous correlation of waves may also be caused by irregularities of the surface and blockiness of the underlying medium. Therefore, the discontinuities in correlation observed in deep seismic sounding and the local changes in the apparent velocity of the first wave in a group may be explained basically either as local flexures in boundaries or as changes in boundary velocities. These same causes may be the explanation for the formation of secondary waves (diffractions from contacts), cusps on travel-time curves, and so on.

A second group of causes for the distortion of the correlation of first waves in groups is interference between groups, which has been discussed earlier.

The origin of a large number of less-well-behaved waves which do not form compact groups but which propagate uniformly between such groups may be explained in a number of ways. If we exclude the effect of the sedimentary sequence (in the so-called open areas) apparently the principal cases are as follows.

1. A local surface in the crust — nonuniform contacts with differences in wave impedance which give rise to reflections; it is apparent that in the vicinity of the critical angle, the apparent velocity for such waves will be very close to the boundary velocity — the layer velocity for the medium beneath the boundary of the inhomogeneity. For angles of incidence which are much larger than the critical angle, these waves will have an apparent velocity approximately equal to the maximum layer velocity in the overlying sequence; that is, if velocity increases with depth, it will be the layer velocity in the layer over the given inhomogeneity;

2. intermediate reflections and better-behaved first waves in groups from irregularly distributed reflecting surfaces; and

3. random scattered vibrations which result in a pseudo-regular wave that is poorly behaved with respect to wave form and correlatability.

Thus, if we assume that the first wave in a group corresponds to a boundary of a reasonably thick (> 5 km) and extensive velocity layer, and that discontinuities in the correlation of this wave result from irregularities in the upper part of this same layer, then most of the following waves in the group and a significant part of the intermediate waves may also be related to discontinuous local irregularities in the crust.

It may be supposed that in such a medium, in which the principal refracting and reflecting surfaces are discontinuous in form and differ only in dimensions — extent of areas with well-defined reflecting and refracting characteristics — the formation of the entire order of multiple waves is less than in a medium with continuous layering. In this respect, in such a discontinuous medium, the primary waves may generally carry more weight than the secondary multiple events. With this concept, it will be of interest to analyze the discontinuity in correlation not only for the first waves in groups, but also for the entire sequence of seismic waves, keeping in mind that the properties of the crust not only change vertically and horizontally between major blocks, but also change over smaller areas along a profile and may reflect small-scale inhomogeneities in the crust.

The intervals over which the first waves in groups can be correlated were listed in Table 12. These are significantly different in different areas even for such dominant events as less-than-critical angle reflections from the Moho. In active regions, these intervals are smaller than in shield areas, where there is no strong noise level, and correlations are less well defined than in depressions, which are characterized by a large number of multiples. These observations provide the basis for supposing that discontinuities in seismic wave propagation reflect discontinuities in the properties of the medium.

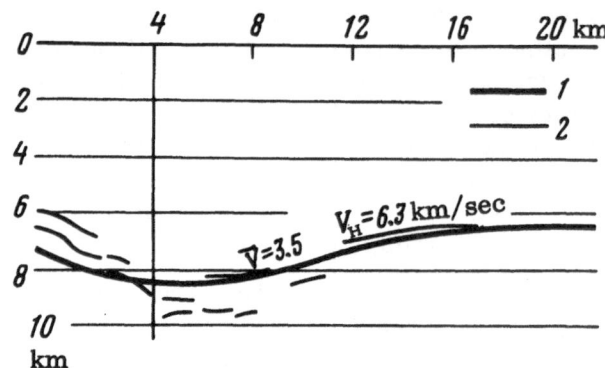

Fig. 4.28. Crustal section constructed with consideration of discontinuities in the correlation of first waves in groups: 1) from averaged travel-time curves; 2) from phase travel-time curves.

Areas where groups of waves generally would not be formed, but the seismic waves would be characterized by combinations of individual events which may be correlated only over short distances are of particular interest. However, in this case, the general character of distribution of the segments of travel-time curves for these waves as a function of time t and distance R is analogous to the behavior characterizing a multiple-layered medium in which velocity increases with depth.

Such areas of poor correlatability in seismic waves are met on many profiles. In some cases, only a portion of the data behaves in this way, only for some specified time interval. For example, on the Russian Platform in the vicinity of the Tatar uplift, the regularity of the P_1^K wave group, related to an intermediate boundary in the crust with a velocity of 7.0 km/sec, is broken (Pomerantseva [131]). On a deep seismic sounding profile across the major Caucasian graben, the P_0^K and $P_{0\,refl}^M$ events are poorly represented by short wave groups and P_i^K waves from intermediate boundaries in the crust are entirely missing [187]. Generalized zones with a high overall level of intensity and with a large number of short, individual events which do not form groups have been noted in the central Caucasus (Antonenko [8]). The interval over which events can be traced is 3 to 10 km on the average, while in the Dzhezkazgan disturbed area (Popov [134]; see also this volume Table 12 and Fig. 4.1), several dominant and well-defined reflected waves may be recognized providing evidence for nearly continuous layering.

It is not yet entirely clear how to interpret discontinuities in seismic wave behavior, or how to distinguish discontinuities in the correlation of waves caused by physical and structural inhomogeneities in the crust from distortions of records caused by secondary waves. However, it is obvious that the present deep seismic sounding sections must reflect not only those features of the velocity profile which have been smoothed out in a general way as a consequence of interpreting grouped travel-time curves, but also the degree of nonuniformity of the entire sequence of seismic waves.

The first efforts in this direction were undertaken as early as 1952.[†] A section constructed on the basis of the first waves in groups corresponding to the surface of the crystalline crust is shown in Fig. 4.28 for an interpretation of deep seismic sounding data obtained in southwestern Turkmenia.

[†] G. A. Gamburtsev et al., "Report on deep seismic sounding efforts along the trace of the main Turkmenian channel," Fondi Inst. Fiz. Zemli, Akad. Nauk SSSR (1952).

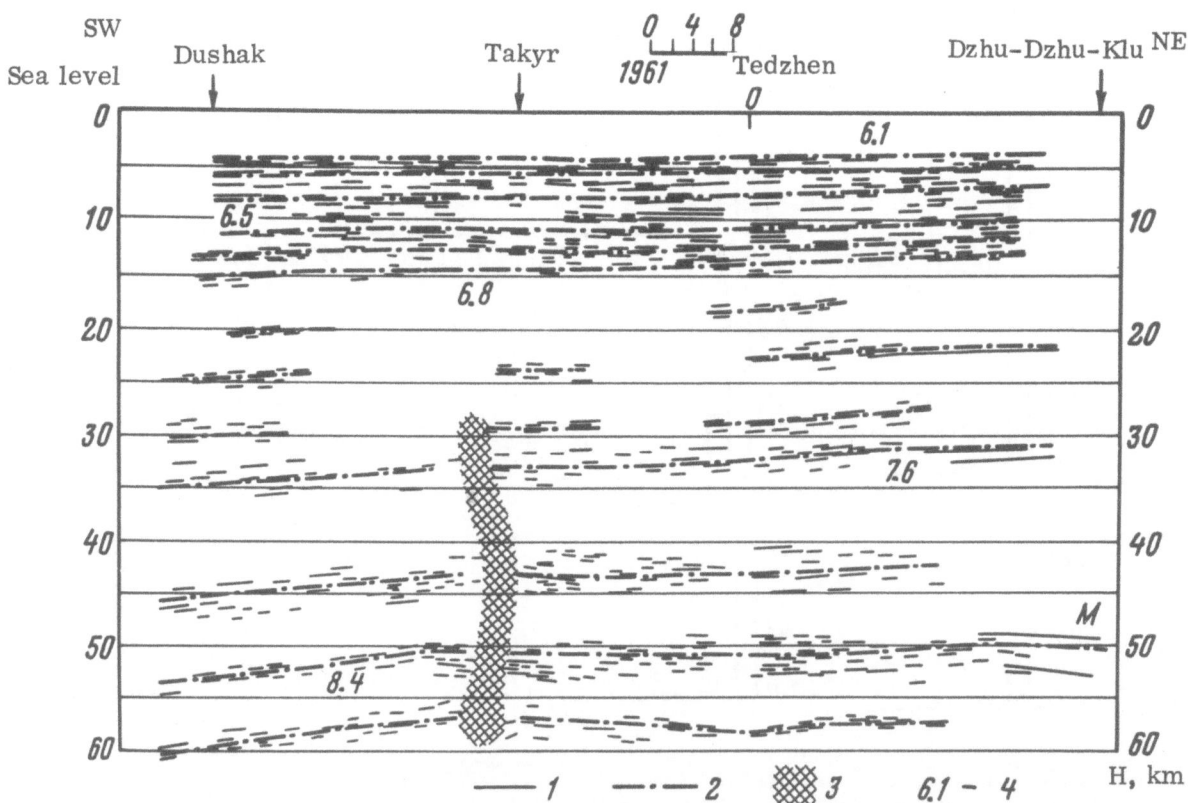

Fig. 4.29. Crustal section along the deep seismic sounding profile from Dzhu-Dzhu-Klu to Dushak (Turkmenia) from data provided by reflections at less-than-critical angles of incidence (Belousov, Vol'vovskii, et al. [20]): 1) reflecting surfaces; 2) conditional reflecting surfaces; 3) zone of anomalous attenuation and complicated seismic records possibly related to deep fractures; 4) boundary velocities (km/sec) from refraction studies.

However, later, as a result of the use of all of the data for the principal wave groups, the basic layering in the section was substantiated while the structural discontinuities were not.

Attempts to consider the intermittent character of travel-time curves were undertaken in relation to the use of low-angle reflected waves, which exhibit a short interval for correlation (Fig. 4.29).

Statistical Evaluation of Discontinuities in Seismic Wave Propagation. In order to evaluate the degree of discontinuity in seismic wave propagation which may occur in a continuously layered medium as well as for comparing the constancy of wave propagation effects observed in different regions, it will be of value to make a statistical evaluation of the correlation ranges for all recorded seismic events. In so doing, we may use distribution plots for the interval over which waves may be traced continuously, ΔL. Similar plots may also be constructed for the first waves in groups to supplement Table 12.

If we assume that discontinuities in correlation are related to discontinuities in boundaries, comparison of the plots for different regions makes it possible to evaluate the representativeness of first waves in groups in comparison with the other waves. The magnitudes of the average values for ΔL may also provide some idea of the average dimensions of possible

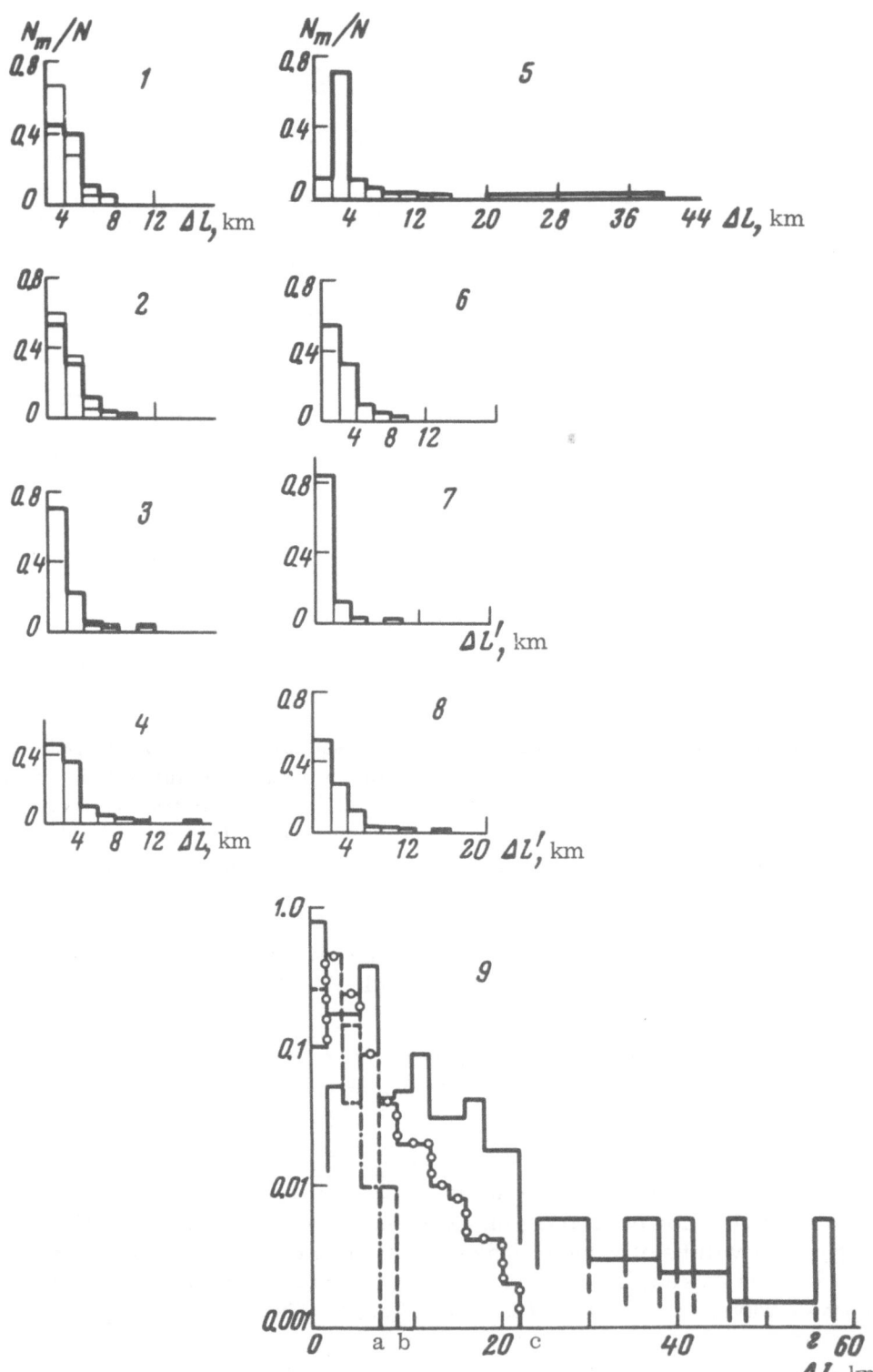

inhomogeneities in the crust either for the areas related with the upper surface of a layer (first waves of groups) or even for inhomogeneities within a layer. This brings up the matter of the "thickness" of these inhomogeneities, which may be evaluated roughly from the average wavelength.

Distribution curves for correlation ranges ΔL for various waves recognized on deep seismic sounding records are shown on Fig. 4.30 for observations in various regions. The plots were based on travel-time curves with consideration of the total number N of wave correlations with a range of more than 0.5 km, that could be traced over the time range 5-15 sec following the first arrival. The collection of plots contains data on the distribution of values for ΔL for various distances R from the shot point (R = 60 km to R = 280 km) and for various intervals for continuity of the data (from 20 to 170 km). Plots 1 and 2 were based on data from crossed profiles.

In addition to the distribution plots for ΔL from travel-time curves, plots 7 and 8 show the lengths $\Delta L'$ of reflecting segments for reflections at greater-than-critical angles.

It is apparent from these data that short intervals of correlation are most common when the behavior of all seismic events is considered; the greatest number of correlative events have ranges of 4-8 km in platform areas and 2-4 km in more active areas.

Long correlative events with ranges of tens of kilometers comprise a larger fraction of the total in platform areas than in more active areas; the distribution plots for platform areas are skewed more toward large values for ΔL (Fig. 4.30, plot 9). The average range ΔL in platform areas is about equal to the correlation range for the first waves in the primary groups (see Table 12).

In active areas, the first waves in groups have the same distribution plots as the rest of the seismic events, and moreover, there is essentially no difference between the plots for transverse and longitudinal profiles.

Fig. 4.30. Distributions of correlation ranges for various waves: 1, 2) southwestern Turkmenia [95], crossed profiles; 3) Predkopetdag basin [42]; 4) northern Tien-Shan, Kurty profile [50]; 5) Ukrainian shield [184]; reflections past the critical angle; 6) Bukhara region (Yaroshevskaya and others, Inst. Fiz. Zemli, 1964-1965); 7) lengths of reflecting segments on section 4.20 [20]; 8) the same, from the section Saatli−Aksu, Azerbaijan [140]; 9) curves are: a) Pamir−Alai zone [96]; b) foothills of Kirghiz graben [50]; c) southeastern Russian Platform [131]; d) Turkmenian Platform, profile from Kara Bogaz Gol to Karashor [69] (last plots are to semilog coordinates).

Region number	N	L	R	Region number	N	L	R
1 (longitudinal profile)	233	31	77	5	165	100	
1 (nonlongitudinal profile)	163	20	77	6	347	68	
2 (longitudinal profile)	173	20	176	7	578	100	
				8	162	70	
2 (nonlongitudinal profile)	267	31	176	9a	384	22	268
				9b	182	40	248
3	160	41	60	9c	473	150	
4	145	48	144	9d	167	170	

N − total number of travel-time segments; N_m − number of segments with a length of ΔL km; R − total length of profile; L − distance from origin of profile; the fine lines indicate nonlongitudinal profiles.

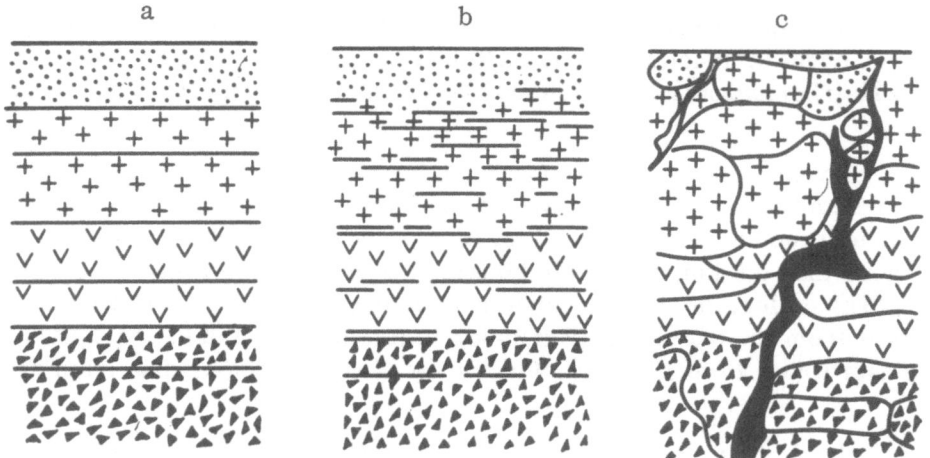

Fig. 4.31. Crustal models [17]: a) continuous layering; b) discontinuous layering; c) macroinhomogeneous blocky layering.

In order to judge whether this distribution for values of ΔL actually corresponds to the multiplicity of vibrations that would be developed in a multiply layered medium, it would be necessary to compare the plots in Fig. 4.30 with similar plots for a theoretically derived set of seismic waves.

If it is assumed, as was done earlier, that most waves are primary and reflect the structure of the medium, then these plots lead to the conclusion that the earth's crust generally is a "macroinhomogeneous" medium with near isotropy in the inhomogeneities that are of moderate size, having dimensions of the order of 2-10 km. Against this background of macroinhomogeneity, the crust appears to be clearly divided into layers, on the basis of velocities. The boundaries between layers in most areas are nearly horizontal or undulate slightly.

The characteristics of the upper parts of a layer for the most part reflect the primary characteristics of the entire layer, but they may vary markedly spatially.

In quiet regions, the dimensions of inhomogeneous regions along the layer boundaries are usually larger than the dimensions of the microinhomogeneities over the entire crustal thickness. This feature is not apparent in active areas. Thus, where no dynamically reflected wave groups are observed, it may be assumed that there is no ordered layering in the crust.

More Complex Crustal Models. With these ideas in mind, there is no limit to concepts of crustal structure either as a multiply layered medium with continuous layering, homogeneous or inhomogeneous. In further refinements of our approximations, it will obviously be necessary to make these models more complex, going first to discontinuously layered models and then to macroinhomogeneous models (Fig. 4.31).

The common feature of all three models is layering — it is reflected by the usual characteristics of seismic wave behavior which we observe from averaged group travel-time curves t(R), and amplitude plots A(R). In addition to these usual characteristics of seismic wave behavior, in deep seismic sounding, we pay attention to statistical summaries of the correlatability of individual waves, whether in groups or between groups, which reflect the discontinuity of boundaries in the models (Fig. 4.31b).

Anomalous behavior of seismic waves in deep seismic sounding, such as diffractions, curvature of travel-time curves, step-like offsets, zones of intense attenuation of waves, and so on, may reflect a blocky medium.

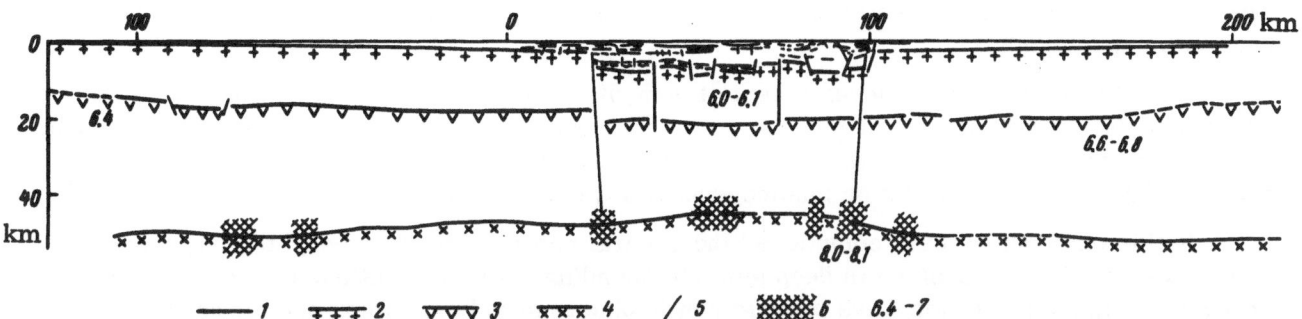

Fig. 4.32. Blocky layered crustal structure based on deep seismic sounding data profile from Zvenigorodka to Novgorod Severskii (Demidenko et al. [66]). 1) Boundaries in sedimentary section; 2) surfaces in the crystalline basement; 3) surface of the "basalt" layer with a velocity of 6.6 km/sec; 4) Mohorovičic discontinuity; 5) discontinuity; 6) zones where the correlation of waves is interrupted; 7) boundary velocities, km/sec.

Usually, in interpreting deep seismic soundings, the major peculiarities in seismic wave behavior are noted as seismic anomalies. An example of such an interpreted section in which a blocky crustal structure is apparent is shown in Fig. 4.32. The blocks are identified on the basis of the behavior of the first waves in the principal wave groups, which, as was seen in Fig. 4.2, are very clearly developed in this region and may be traced reliably over long distances.

In other regions, more complicated seismic anomalies are found which make quantitative interpretation difficult. Therefore even in cases in which seismic anomalies are indicated on a section, some basis is provided for a purely geometrical and intuitive presentation of the geology, though only schematically.

The amount of information provided by a deep seismic sounding section is increased significantly when auxiliary data in the form of reflecting and refracting boundaries, based on an analysis of wave forms of individual waves, are added. Such procedures have been described in [140] and [160].

Inasmuch as the theoretical basis for wave propagation in a discontinuous layered or a blocky layered macroinhomogeneous medium has not yet been well developed, we recommend a wider use of qualitative methods for representing the character of wave behavior on seismic sections, determining the positions of individual "effective" reflecting and refracting surfaces, arrived at using elementary geometric considerations that have been developed for a continuously layered medium.

In order to develop a more satisfactory approach to the interpretation of deep seismic sounding data, it will be necessary to develop the theory for wave propagation to include these more complex media. This would permit a solution to the direct interpretation problem and would allow determination of which parts of the discontinuous behavior in wave propagation correspond to primary inhomogeneities in the medium and which may be explained by secondary interference phenomena.

In concluding this chapter, we should also note some inadequacy in the method of analyzing the normal behavior of seismic waves now used with models having continuous layering. In determining the nature of the observed wave groups, each individual wave is usually considered independently of all the other waves propagating through the medium. Usually, the conditions for their interference with each other are not considered in theoretical analysis. As has been shown in § 4, these effects play an important role even in the case of a single boundary.

The significance of multiple reflections increases in a multiply layered medium, such as we deal with experimentally.

It is obvious that in the development of computer programs for interpreting deep seismic sounding data, we will need first to consider interference effects for all combinations of waves present, and which take part in the development of the seismic wave field for various types of multiple layered homogeneous or inhomogeneous layered media.

In this chapter, we have considered the general nature of wave propagation represented by the function $A(t, R)$ involved in deep seismic sounding with observations being made along profiles. In most areas, the wave propagation is observed to be complex and discontinuous consisting of dominant elements — waves. Treating waves in groups allows systematization of the seismic wave field. It is possible to construct travel-time curves $t(R)$ and amplitude plots $A(R)$ for wave groups just as for individual waves.

These combined wave groups have curves for $t(R)$ and $A(R)$ which behave similarly to the corresponding curves for individual reflected or refracted waves propagating in a layered weakly inhomogeneous medium for either a single or a series of shot points.

In evaluating the general character of wave propagation and the properties of wave groups, we must stress that some errors may be involved in systematizing the wave behavior in terms of correlations of wave groups, in comparison with the results obtained using individual waves, because the procedure is more or less subjective.

The properties of the seismic waves in various regions is determined overall by their dominant elements: the principal wave groups, and the degree of continuity of the wave field. These normally characterize the seismic wave field, which usually does not differ much within the limits of a single tectonic zone and which characterizes the average layering of the crust, subordinating its graininess which accompanies more or less obvious seismic anomalies reflecting the gross blockiness of the crust. Deciphering the normal and anomalous characteristics of a seismic wave field is the primary problem in deep seismic sounding. In view of the fact that many of the characteristics of the seismic wave field are interpreted only qualitatively, it is important to have general ideas of the principal characteristics of the field for various types of crust. This allows classification of an area under study according to the type of seismic wave behavior as the first step in analyzing deep seismic sounding data — during correlation of seismic records and construction of travel-time curves for the primary groups of deep waves.

CHAPTER V

WAVE PROPAGATION IN CRUSTS OF VARIOUS TYPES

The grossly generalized geological and geophysical compilations and summaries of results of seismic investigations of crustal structure in various regions are usually based, if not always, only on the resultant velocity sections obtained in these areas. Each of these has been constructed on the basis of some arbitrary analysis of the observed features of wave propagation. However, we have not yet closely examined this topic. Moreover, the topic is quite obscure.

In fact, the seismic wave propagation picture observed in various areas may depend not only on objective causes, variations in the structure, and properties of the medium, but also on such factors as variations in the type of equipment used, the survey procedures, and interpretation.

Questions about the comparability of interpreted results arise as a consequence of differences in the physical assumptions used in analyzing seismic wave fields. The same set of travel-time curves may be interpreted in different ways (as for example, with and without consideration of intermediate boundaries). There may be a considerable degree of incompatibility, for example, between the ideas about layering in the crust for sections derived from data using different degrees of detail.

It is obvious that in order to have reasonable compatibility between sets of seismic data obtained in different areas it is extremely important to start at the very first stage and particularly to recognize the primary parts of the seismic wave field which are diagnostic for various types of crustal structure, and which may be identified even with different survey procedures and serve as landmark characteristics in comparisons.

Such diagnostic features characterize waves from the sedimentary section (S), the crust (K), and the mantle (M).

Three types of crustal structure may be recognized on the basis of the velocities and wave forms for these waves: continental, oceanic, and transitional. The last type is sometimes further divided into near-oceanic and near-continental. We will review briefly the characteristics of seismic wave propagation in these types of crust (this subject has also been covered in [3, 103, 163]).

§1. Characteristics of Wave Propagation for the

Principal Crustal Types

In comparing records obtained with equipment having the same frequency response characteristics in areas characterized by continental and oceanic types of crust, a simple form is noted for oceanic records and a complex form for continental (Fig. 5.1). The simple form for oceanic records is explained by the presence of a water layer in which shear waves do not

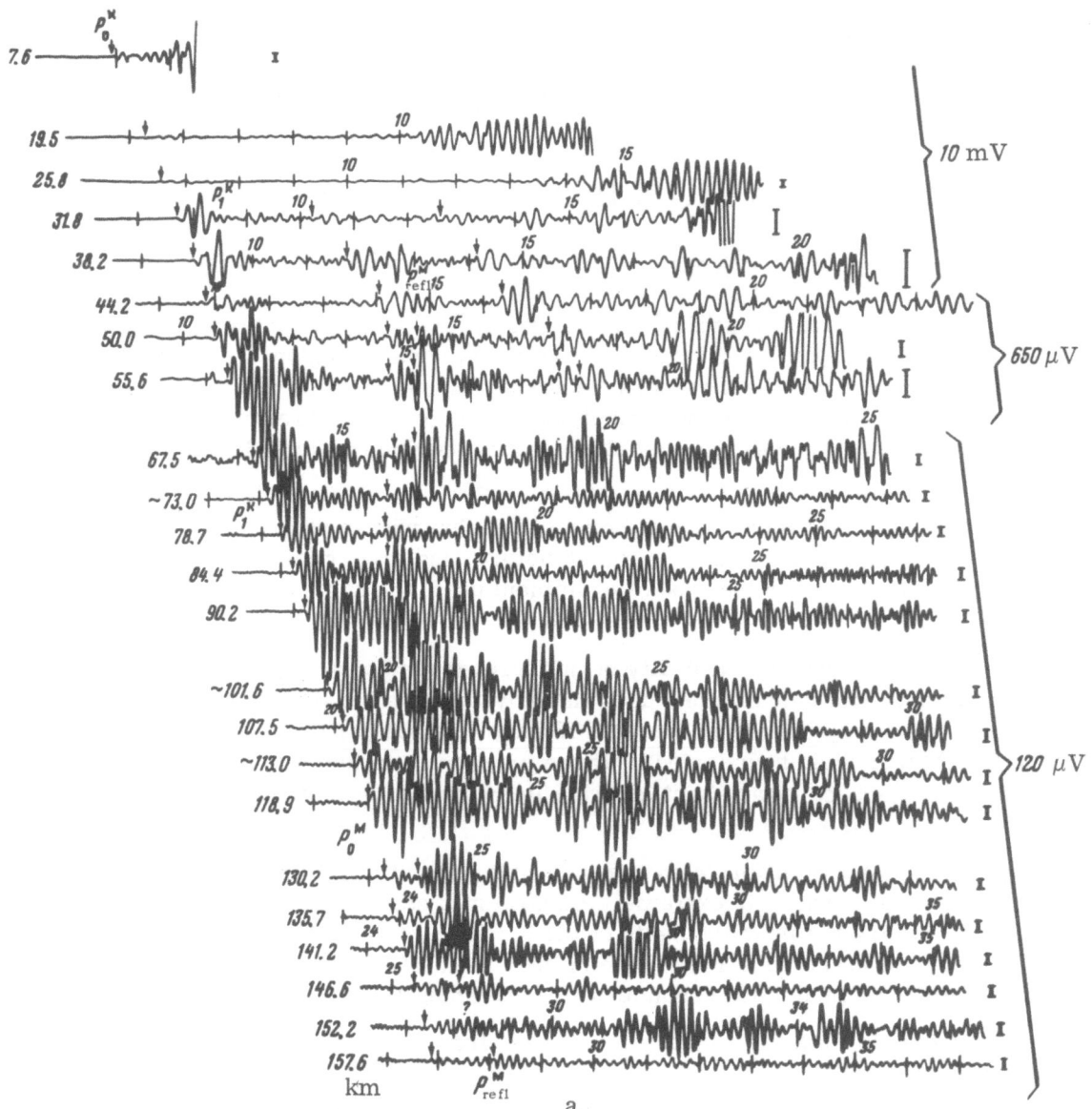

Fig. 5.1. Deep seismic sounding records for two different types of crust: a) continental Kuriles (amplitude-controlled signal on the right). These records, made over a fre- shown in Figs. 4.2-4.4 (Zverev et al., 1964).

propagate and, apparently, to a simpler crustal structure, especially in the upper part of the crust [108; 163, Chapter 2]. Under similar conditions on land, a variety of multiples and shear waves are generated which give rise to complicated coherent waves and strong incoherent noise in the later parts of records.

Keeping these general differences between wave propagation for the continents and oceans in mind, we will compare the deep traveling events, considering that they are the most significant.

The primary deep traveling waves may be identified most satisfactorily with various survey procedures either when they may be traced as the first arrival, or when these waves are dominant in later parts of the record and may be easily recognized with respect to amplitude and wave form even in the presence of noise.

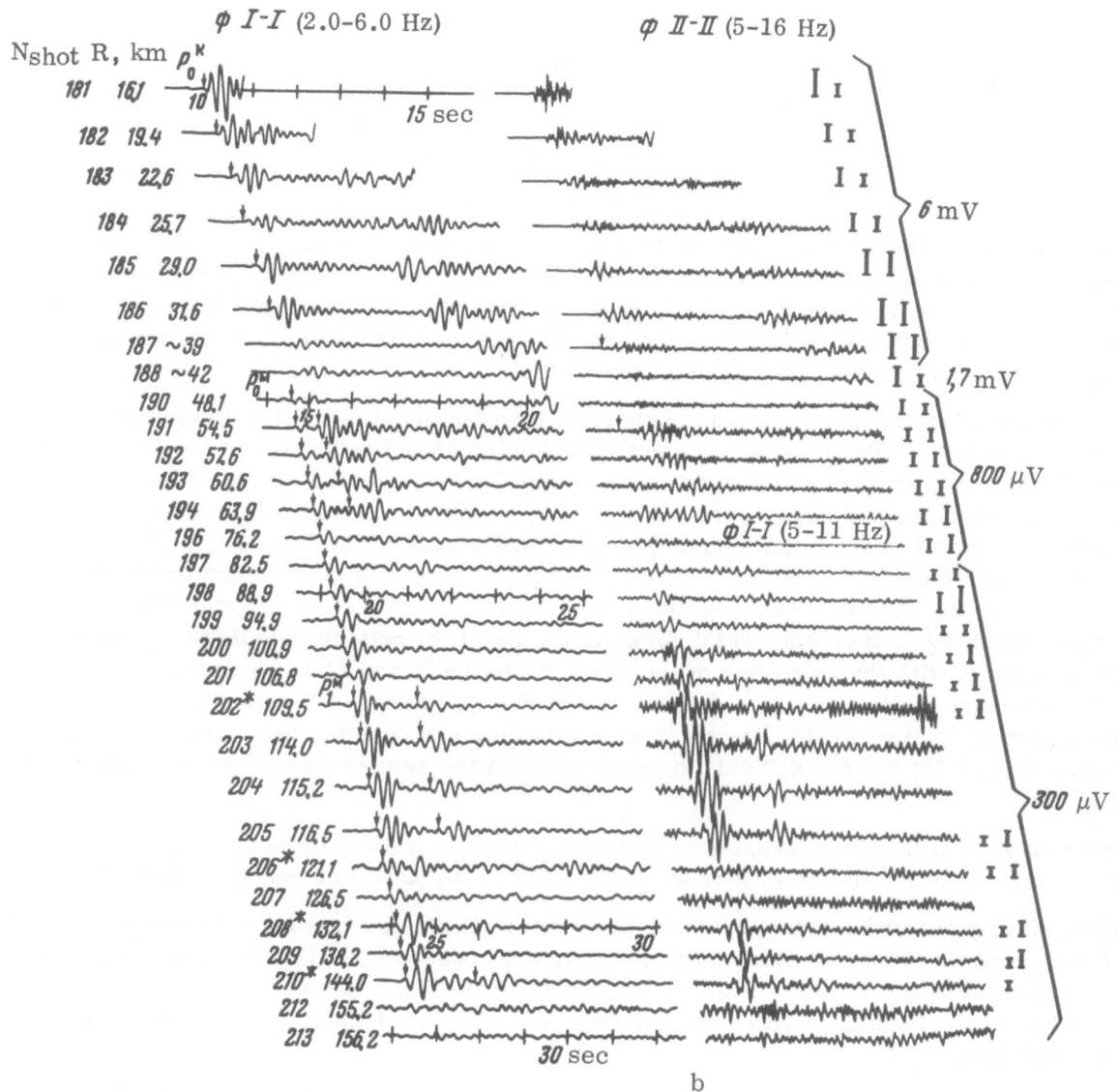

b

crust, Sakhalin area (Zverev et al., 1963-1964); b) oceanic crust, east of the southern quency range of 5-16 Hz, may also be compared with the records for a continental crust

Waves of this type include the P_0^K event, corresponding to the surface of the crystalline part of the crust, which emerges as the first arrival ahead of the sedimentary events (Fig. 5.2) and the wave P_0^M, which overtakes the P_i^K event. A wave which is dominant as later arrival is the $P_{0\,refl}^M$ event — a reflection from the M-discontinuity (preferentially, at a less than critical angle).

The group of waves P_i^K corresponding to intermediate boundaries in the crust may not always be identified as first arrivals. In many cases, they are not dominant, and so, may not be recognized at all.

The differences in seismic wave behavior between continental and oceanic types of crust may clearly be traced through data obtained in the transitional zone between Asia and the Pacific Ocean (see [163]).

The velocities for the deep-traveling waves P_0^K and P_0^M are listed below [22]:

	Continental crust	Oceanic crust
Surface of the crystalline part of the crust (P_0^K)	5.2—6.4 km/sec,	6.4—6.8 km/sec,
M-discontinuity (P_0^M)	7.5—8.2 km/sec,	7.9—9.4 km/sec.

These data, as did those presented earlier in Chapters III and IV, indicate that the essential differences in wave forms for continental and oceanic crusts consist of different velocity characteristics for the events related to the surface of the crystalline part of the crust. For a continental crust, this event is usually a wave with a velocity of about 6 km/sec (however, there are a number of exceptions in the case of crust beneath basins and inland seas [58, 160]); for an oceanic crust, this event has a velocity of 6.5-6.8 km/sec.

A second distinctive feature of wave behavior for a continental crust is the presence of clear and dominant reflections at less-than-critical angles from boundaries in the lower part of the crust and the upper part of the mantle, including the M-discontinuity (see Figs. 4.2 and 4.4). In areas with an oceanic crust, usually high-angle reflections cannot be recognized from boundaries in the crust or from the Moho. The principal events in the later parts of records in such a case are strong body waves of the type $PS_0^K P$ (PS*P on Fig. 5.2), which propagate as shear waves along the boundary between the sediments and the crystalline part of the crust and multiple waves $P_0^K{}_{KP}$ and $P^M{}_{KP}$ which form as the result of reflection within the water layer and propagate with the velocities of the deep traveling ^{I}K and M waves [163]. With respect to high-angle reflections from boundaries in the upper part of the mantle, it has been indicated by recent studies (S. M. Zverev et al.)[†] that they are possibly present on records as early events (Fig. 5.1b) but are difficult to separate from the earlier waves at low frequencies (about 5 Hz).

In the case of transitional crustal sections, the seismic wave field is characterized by the same general features as either a continental crust or an oceanic crust [101, 163].

Thus, in regions with a near-oceanic type of crust, the seismic wave behavior is similar to that for an oceanic crust, but it is complicated by the presence of a thin layer of sediments. Also, the $PS_0^K P$ waves are absent. This is apparently a result of a marked change in the velocity at the surface of the crystalline part of the crust, and to a lesser degree of homogeneity of this surface [24].

In the case of a near-continental type of crust in which the velocity characteristics are close to those for a continental crust, reflections at less-than-critical angles from the M-discontinuity are less well developed or completely absent, and waves from intermediate boundaries in the crust cannot be easily resolved. The principal events in such areas are related to boundaries in the lower crust and with the M-discontinuity. The M-discontinuity is not clear and is usually difficult to recognize. Studies by Yu. V. Tulina[‡] in the southern Kuriles region in areas with a near-continental type crust have shown that in many cases waves with a velocity of 7.2-7.4 km/sec, or lower than the velocity for the M-discontinuity in areas with a typical continental crust, can be traced at large distances from the shot point (more than 150 km). Analysis of such records at various frequencies provides a basis for assuming that in such areas layers change in thickness and the crust has a blocky structure, which may explain the

---

[†] S. M. Zverev, Yu. V. Tulina, and others, "Study of crustal structure in areas of the southern and central parts of the Kuriles trench and southern Sakhalin using the deep seismic sounding method." Report of operations during 1963-1964, Fondi Inst. Fiz. Zemli (1966).

[‡] Yu. V. Tulina, "Detailed seismic study in the southern Kuriles Archipelago," Proc. of the Far-Eastern Summary Session of the ONZ, Acad. Sciences, USSR (1965).

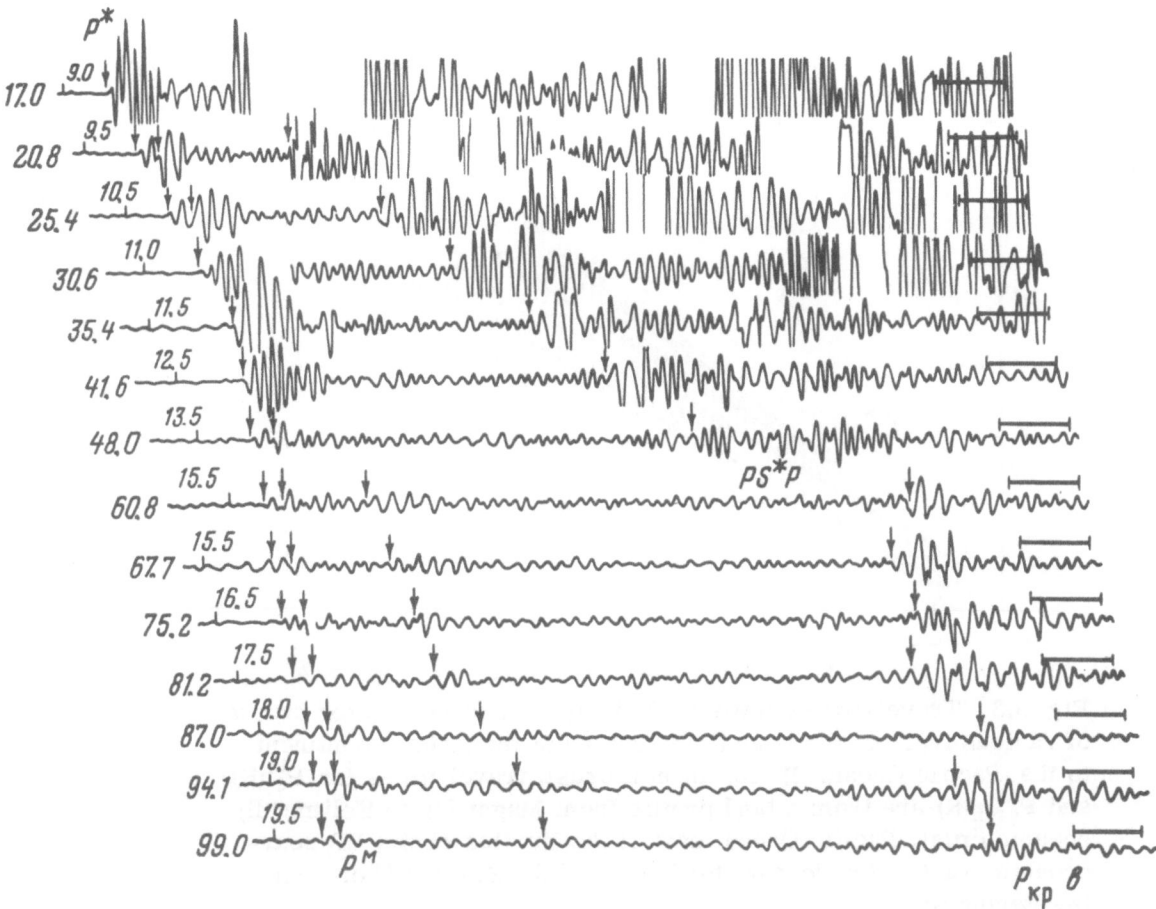

Fig. 5.2. Deep seismic sounding records from a transitional region between Asia
and the Pacific Ocean [33].

marked changes in character of seismic waves over moderate distances along a traverse (over
several tens of kilometers). It should be noted that the nature of the seismic wave field depends
on the orientation of the deep seismic sounding profiles relative to the strike of the island chain.
Considering this, we may also speak of the structural anisotropy of seismic wave behavior.
Such effects have not yet been adequately studied. However, it is clear that such a behavior is
one of the basic forms, and we have not included it in the classification scheme with which we
have characterized the diagnostic features of continental and oceanic wave fields, observed
over large areas of the continents and oceans.

We evaluated the normal behavior of seismic wave propagation in Chapter IV. Such
behavior reflected the layering in a medium, that is, the change in velocity with depth and the
persistence of a section over a region. We might suppose that there is a similar behavioral
pattern for seismic waves related to a blocky structure, superimposed on the background of
the layering pattern. When wave propagation effects are anisotropic, blockiness dominates,
and as a result, the normal wave patterns caused by layering are unclear and change as a
function of the orientation of the seismic profiles.

The phenomenon of structural anisotropy in wave behavior will obviously be observed
with reasonably detailed studies of other linear structural features, as, for example, the
Caucasus, but apparently the effect is more obvious for island arcs and midocean ridges,
separating areas with oceanic and near-oceanic crustal structures.

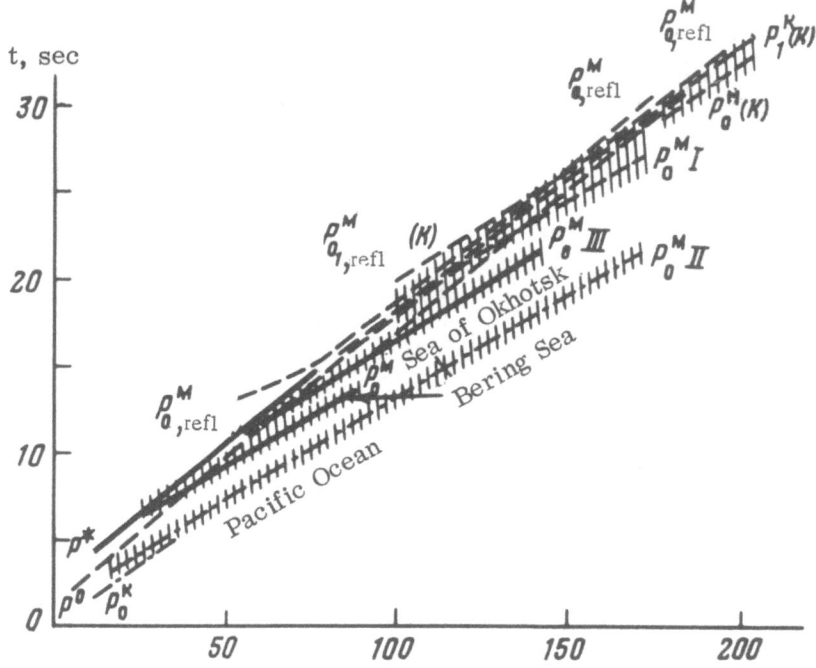

Fig. 5.3. Travel-time curves for principal wave groups for crusts of various types in the transition zone from the Asiatic continent to the Pacific Ocean: I) continental crust, travel times for P_0^M(K) and $P_{0\text{ refl}}^M$(K) are from a land profile from Magandan to Kolima; II) oceanic crust, Pacific Ocean, east of the Kuriles chain; III) near-oceanic crust in the deep-water basins of the Sea of Okhotsk and the Bering Sea.

Travel-Time Curves for Crusts of Various Types. The quantitative expressions for the various seismic wave fields for various types of crust differ with respect to the travel time for the principal groups of deep traveling waves recorded at equivalent distances from the source (Fig. 5.3). For an oceanic type of crust, the travel-time curves have two segments, with velocities of 6.5-6.8 km/sec and more than 8 km/sec. Travel-time curves for a continental type of crust have several segments, and among the first parts of the P_i^K waves there is nearly always a segment for the P_0^K event with a velocity close to 6 km/sec (Fig. 5.3). The travel-time curves for the first waves show a larger travel time for a continental crust and a lesser travel time for an oceanic crust. This is readily apparent from Fig. 5.3 or from a comparison of the travel times for the first arrivals at equal distances from the source (see below, Table 21). In this table, the oceanic travel times have been corrected for the ocean depth. Parenthesis identify areas for the recording mantle events P_i^M. A comparison of travel times for the P_0^M event, related to the M-discontinuity, allows us to evaluate the differences in crustal thickness, if we assume that the average velocity in the crust is constant. This is reasonable for a crude estimate if we exclude areas with markedly different thicknesses of sedimentary rock with low velocity.

Wave Forms for the Principal Wave Groups with Different Types of Crust. The nature of changes in the relative intensity of the principal wave groups for oceanic and continental types of crust is represented by the dynamic travel-time curves shown in Fig. 5.4. These illustrate the dominant nature of the reflection from the M-discontinuity in the case of a continental crust, and the dominance of the first arrival P_0^M on records obtained in oceanic and near-oceanic areas.

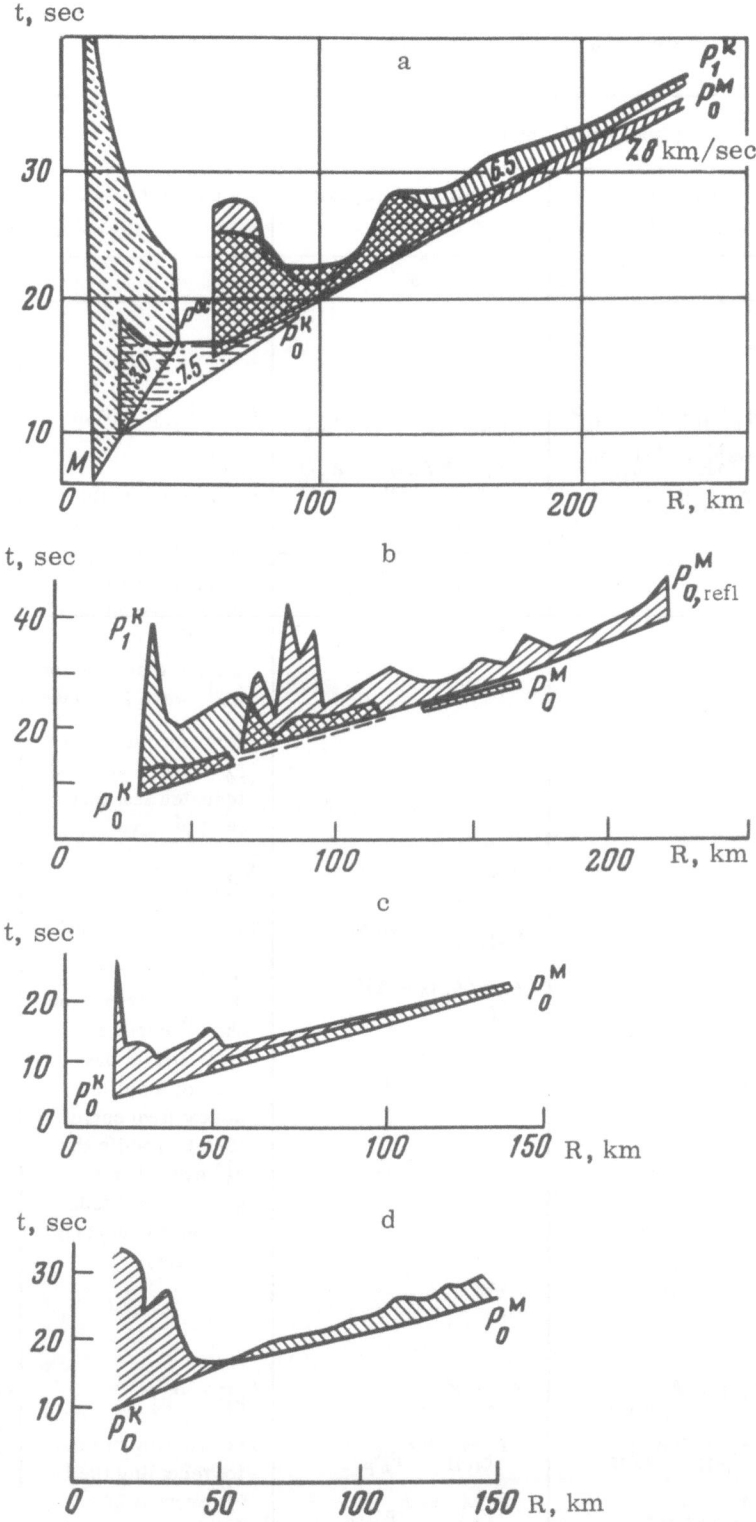

Fig. 5.4. Dynamic travel-time curves for various types of crust: a) continental crust, Caspian Sea—Turkmenian Platform [3]; b) the same, northern part of the Sea of Okhotsk—Okhotsk Rise [33]; c) near-oceanic crust, south Kuriles basin of the Sea of Okhotsk; d) oceanic crust, Pacific Ocean southeast of Urup Island (Zverev et al., 1964).

TABLE 19. Summary of Dynamic Characteristics of Seismic Waves

Type of crust and section*	Amplitudes of the most obvious events		Observed wave characteristics	Type of observed waves
	uniformly layered model	nonuniformly layered model		
1	2	3	4	5

Oceanic
 	
km   km/sec
$H = 5$
$h^B = 5$ $V^B = 1.45$
$h^{sed} = 1$ $V^{sed} = 2.0$
$h^K = 8$ $V_0^K = 6.3 \times$
      $(1 + 0.02 h_0^K)$
      $V^M = 8.0$
	$A_{P_0^M\,refl} > A_{P_0^K\,refl} >$
	$A_{P_0^M\,H} > A_{P_0^K\,H} >$
	$A_{P...S_0^K...P}$
		$A_{P...S_0^K...P}$ $A_{P_0^M\,refl} \approx$
		$A_{P_0^K\,P} \gtrsim A_{P_0^M\,P} \gg A_{P_0^M\,H}$
		
		Limiting distance for recording $P_0^K P$ and $P...S_0^K{}_P...P$ events is 85 km
		
		$A_{P_0^K\,P(KP)} \approx A_{P_0^K}$
		$A_{P_0^M\,P(KP)} \approx A_{P_0^M\,P}$
		$A_{P_P^K} / A_{P_H^M} = 100$
		$A_{P_P M} \approx A_{P_0^K\,P}$
			$A_{P...S_0^K...P} \gtrsim A_{P_0^K} > A_{P M}$
			
			In some areas, the limiting distance for recording the P_0^K event is about 100 km. Attenuation of the P_0^K wave is weaker than that for a P_H^K wave in a uniform medium. The P_0^K wave is attenuated less than the P_0^M wave
			
			$A_{P_0^K\,(KP)} \gtrsim A_{P_0^K}$
			The $P_0^M\,P(KP)$ wave is not clearly recognizable. Sometimes the P_0^K event as a later arrival has a greater amplitude at low frequencies (3–6 Hz) while the P_0^M event has a greater amplitude at high frequencies (6–17 Hz)
				P_0^K-refracted or multiple refracted wave in the crystalline part of the crust
				P_0^M-head waves or weakly refracted waves
				
				$P...S_0^K{}_P...P$ event is the resultant of a transverse wave in the crystalline part of the crust and a body wave reflected from the M-discontinuity
				The P_{KP}^K event is a multiple reflected and refracted wave in the water layer

Suboceanic km km/sec $H = 17$ $h^B = 2$ $V^B = 1.45$ $h^{sed} = 5$ $V^{sed} =$ $2(1 + 0.15 h^{sed})$	$A_{P_0}^M > A_{P_0\,refl}^K >$ $A_{P_0\,H}^M \geq A_{P_0\,H}^K \gg$ $A_{P...S_0^K...P}$	$A_{P...S_0^K...P} \approx$ $A_{P\,refl}^M \approx A_{P_0\,P}^K \approx$ $A_{P_0\,P}^M \gg A_{P_0\,H}^M$	$A_{P_0}^K > A_{P_0}^M$ The limiting distance for recording the P_0^K event is 80–100 km	P_0^K-refracted wave P_0^M-head wave, or weakly refracted wave

*H) Depth to M-discontinuity; h_B) ocean depth; h^{sed}) thickness of sedimentary section; $h_{0,1}^K$) thickness of layers in the crystalline part of the crust.

TABLE 19 (continued)

Type of crust and section*	Amplitudes for the most obvious events		Observed wave characteristics	Type of observed waves
	uniformly layered model	nonuniformly layered model		
1	2	3	4	5
$h^K = 10$ $V_0^K = 6.3 \times (1 + 0.008 h_0^K)$ $V^M = 8.0$		$A_{P_0^K P}/A_{P_0^M H} = 10$ The limiting distance for recording the $P_1^K P$ event is 110 km $A_{P_{P(KP)}^K} < A_{P_P^K}$	The P_1^K event is attenuated less than the $P_1^K H$ event, as well as the P_0^M event $A_{P_{1(KP)}^K} \geq A_{P_1^K}$	P_1^K(KP)-multiple reflected-refracted wave from the crystalline part of the crust and the earth's surface
Continental I km km/sec H = 17 $h^B = 1$ $V^B = 1.45$ $h^{sed} = 1$ $V^{sed} = 2.5$ $h_0^K = 3$ $V_0^K = 5.9 \times (1 + 0.01 h_0^K)$ $h_1^K = 12$ $V_1^K = 6.5(1 + 0.004 h_1^K)$ $V_0^M = 8.0$	$A_{P_0^M refl} > A_{P_0^K refl} >$ $A_{P_0^M H} > A_{P_0^K H} >$ $A_{P_0^0 H}$	$A_{P_0^0 P} \approx A_{P_0^K refl} \approx$ $A_{P_0^K P} \gg A_{P_0^M H}$ $A_{P_0^K P}/A_{P_0^M H} = 20$ The limiting distance for recording $P_0^K P$ and $P_0^M refl$ waves is 160 km	$A_{P_0^M refl} \geq A_{P_0^K} \approx$ $A_{P_0^M} \approx A_{P_0^K}$ Attenuation of $A_{P_0^K}$ and $A_{P_0^M}$ waves is less than that of head waves $A_{P_{0(KP)}^K} \approx A_{P_0^K}$ $A_{P_{1(KP)}^K} \approx A_{P_1^K}$	P_0^K-head wave P_1^K-weakly refracted wave F_0^M-weakly refracted wave $P_0^M refl$-combination $(P_{refl}^M + P_{1 refl}^K)$ $P_{0(KP)}^M$ multiple reflected wave in sedimentary section
Continental II km km/sec H = 31 $h^{sed} = 1$ $V^{sed} = 3.5$ $h_0^K = 5$ $V_0^K = 5.1 \times (1 + 0.02 h_0^K)$ $h_1^K = 10$ $V_1^K = 5.9 \times (1 + 0.008 h_1^K)$ $h_2^K = 3$ $V_2^K = 6.37$ $h_3^K = 12$ $V_3^K = 7.07$ $V_0^M = 8.0$	$A_{P_0^M refl} > A_{P_1^K refl} >$ $A_{P_0^M H} > A_{P_1^K H} >$ $A_{P_0^K H}$	$A_{P_0^K P} \approx A_{P_0^M refl} \approx$ $A_{P_0^K refl} > A_{P_0^M H} >$ $A_{P_1^K H}$ $A_{P_0^M refl}/A_{P_0^M H} > 20$ The limiting distance for $P_1^K P$ is 120 km $A_{P_1^K P}/A_{P_0^M refl} \approx 2\text{-}3$	$A_{P_0^K} \approx A_{P_0^M} >$ $A_{P_0^K} > A_{P_0^M}$ $A_{P_0^M refl}/A_{P_0^M} > 5$ Attenuation of the $P_0^M refl$ wave is less than in an ideally elastic medium P_0^K attenuated less rapidly than $P_0^K H$ and has a well-defined record form at low frequencies at distances up to 50-70 km	P_0^K-refracted wave P_1^K-(observed characteristics uncertain) P_1^M-weakly refracted wave $P_0^M refl$-reflection from the M-discontinuity

<u>Concerning the Nature of Waves and Type Crustal Models.</u> Following the usual logical ideas accepted in interpreting deep seismic sounding data (see Chapter IV), the nature of waves and models of the medium for various types of crust in the transitional zone are determined by comparison of the experimental data with theoretical calculations. A seismic section constructed on the basis of geometrical concepts and the simplifying assumption that the travel times for the first waves in the groups correspond to simple wave fronts associated with boundaries characterized by velocity discontinuities, while refracted waves are assumed to be head waves, is basic for such calculations. Travel-time curves and amplitude plots have been computed for models consisting of uniformly layered sections. Then, nonuniformities are added to these sections. In so doing, the section is modified in such a way that the wave forms and amplitudes for the first waves in groups remain close to the observed data. The results are a collection of travel-time curves and amplitude plots for nonuniformly layered sections of oceanic, near-oceanic, and continental types of crust, with the characteristics listed in Table 19.

The relative amplitudes of the primary elements in the wave field — longitudinal waves which can be traced in the early part of the record and some body waves which are characteristic for oceanic and near-oceanic models — are described in Table 19 for four models.

Columns 2 and 3 in this table give the theoretically expected relations (for ideally elastic media), column 4 gives the experimentally observed relations, and column 5 lists the conclusions which may be drawn about types of waves.

We will now analyze the data presented in the table from the point of view of evaluating the medium under study. Consider the influence of the following factors:

1. The number of seismic events which are recognized is much less than the number needed from the point of view of dynamic probabilities. Thus, in most areas two (in the case of a near-oceanic crust), three (oceanic), or four (continental) wave groups are recognized, while the theoretical models were evaluated on the basis of 5 or 6 groups.

The P_{refl}^{K} wave, which apparently is present in areas of continental crust in the group $P_{1,2...}^{K}$ and cannot be distinguished at the frequencies used is consistently not observed. For the same reason, apparently, the $P_{0\,refl}^{M}$ wave is not recognized on records from areas with an oceanic or near-oceanic crust, where because of velocity relationships, these waves cannot be distinguished from P_{0}^{K} waves, and in cases where there is a velocity gradient in the crust, they usually have a limiting distance for observation.

The $P_{0\,refl}^{M}$ waves observed in areas of continental crust are dominant in amplitude, so these have been evaluated in calculations for both uniform and nonuniform media.

2. Attenuation of observed refracted waves in most cases is less than the attenuation for head waves in an ideally elastic medium. The P_{0}^{K} waves have a well-defined limiting distance for observation in zones with a near-oceanic crust and in some areas with an oceanic crust. Thus, it may be concluded that in most cases, the P_{0}^{K} wave is the combination of a refracted and reflected event. In cases when they are traced as later arrivals, they may be multiple refractions at the greater distances.

3. In regions with a continental crust, the P_{0}^{M} waves attenuate less rapidly than head waves, and so they may be taken to be refracted waves.

In regions with an oceanic crust, the P_{0}^{M} waves are apparently more complex and consist of the superposition of a head wave and a reflected wave at a less-than-critical angle associated with the upper part of the subcrustal layer, and a refracted wave propagating in the upper part of a region with a very slight velocity gradient.

4.   The $P...S_{0P}^K$ event is a very complex body wave related to the refracted wave S_{0P}^K in the basaltic layer of an oceanic crust and with reflection of a dilatational wave from the M-discontinuity. These waves have very similar travel-time curves and about the same amplitudes, which may explain their large amplitudes on records. A simple dilatational wave in a uniformly layered medium should be only half as strong as the P_0^K wave, as indicated in column 2.

5.   The observed relative amplitudes are in good agreement with the values calculated for a model of a nonuniformly layered medium despite the fact that they do not agree quantitatively with theory. The maximum differences between the amplitudes of the observed principal wave groups recorded simultaneously are no more than an order of magnitude, while in the case of the calculated amplitudes, if we assume the P_{0H}^M wave exists, this range amounts to three orders of magnitude for all of the models examined.

Differences of an order of magnitude between the calculated wave intensities are possible for variations of the medium where all of the layers are nonuniform; that is, all of the refracted events are refracted geometrically. The observed amplitudes and travel times for the P_0^M were, especially for an oceanic crust at distances up to 100 km, behave in this way to a greater extent than in the case of head waves.

Thus, the general characteristics of the basic components in the wave field (the dynamic ranges of waves over areas where they are superimposed, the nature of attenuation, the existence of limiting points, and slight curvature of travel-time curves) allow us to assume that in all areas, layers in the earth's crust exhibit velocity inhomogeneities and the velocity increases gradually with depth within the layers.

However, in a few cases, this general behavior may not hold within a single region being studied. Attenuation of the principal wave groups is frequently rapid and sometimes approaches the attenuation expected for head waves. This might be explained in two ways: a higher degree of uniformity of the layers or a greater attenuation in nonuniform layers. The matter of which explanation holds in specific cases can be resolved only by additional experimental work.

Selection of Crustal Models by Computer Programs. A computer program has been written by V. I. Keilis-Borok and T. B. Yanovskaya to select a multiple layered nonuniform model corresponding to an observed set of travel-time curves with a prescribed accuracy. The method used in solving this inverse problem consists of a selection of variations to the velocity profile V(H), which satisfy the observed travel times t(R) over a band of values, t(R) + Δt(R) and limits for the properties of the section. The number of layers, the limits for velocities in the layers, and the limits of depths are provided as output.

In the case of deep seismic sounding data, the band of values for Δt(R) for horizontal boundaries (a program was also written for such an assumption) may be represented, as is evident in Fig. 5.3, by the possible variation in the thickness of the crust caused by changes in the velocity and depths to intermediate boundaries.

The velocity profiles computed with this program may be supplemented with computed curves for the amplitudes of refracted waves.

Computations were carried out by T. B. Yanovskaya for oceanic and near-oceanic types of crust. We will give the results for a near-oceanic type of crust. The travel times are given in Table 20. The band of times taken was Δt = 0.7 sec. The first arrivals in the central part of the southern Kuriles deep-water basin of the Sea of Okhotsk fall in this interval. The limits on depths and velocities are indicated on Fig. 5.5.

Forty-eight velocity profiles were obtained from the calculations which could be divided into two groups: profiles with a positive velocity gradient in the crystalline part of the crust

TABLE 20. Travel Times for a Near-Oceanic Crust

R, km	t, sec	R, km	t, sec	R, km	t, sec
10	7.9	50	13.6	90	18.7
15	8.6	55	14.3	95	19.3
20	9.4	60	15.1	100	19.9
25	10.1	65	15.7	105	20.5
30	10.8	70	16.3	110	21.2
35	11.5	75	16.9	115	21.8
40	12.2	80	17.6	120	22.4
45	12.9	85	18.1	125	23.0

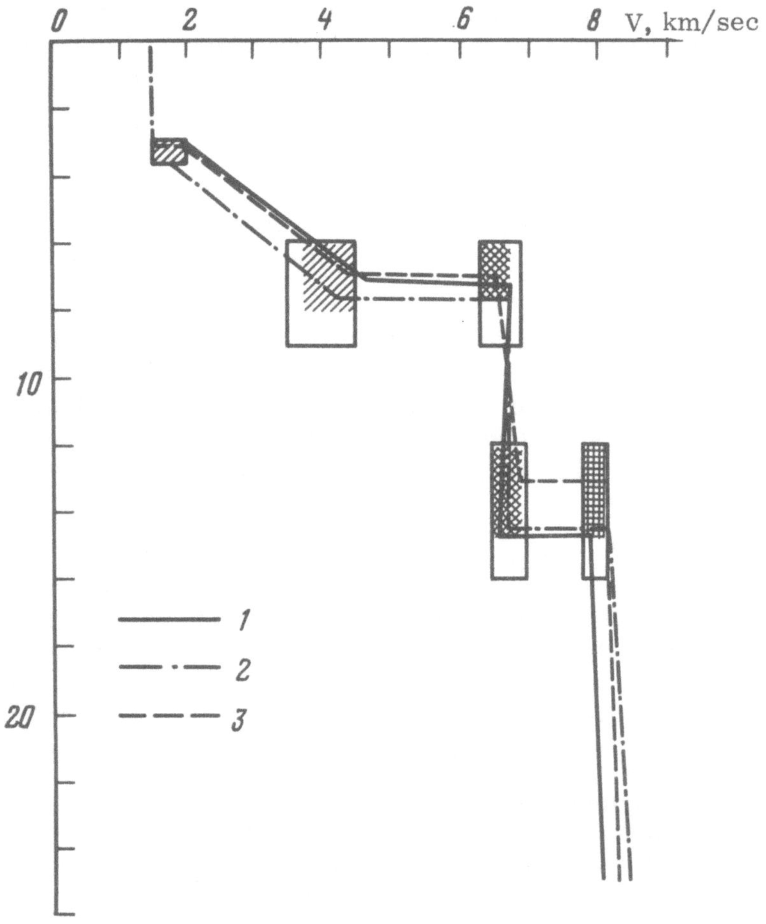

Fig. 5.5. Velocity profiles compiled by solution of the inverse problem (using Yanovskaya's program). Types of sections: 1) with a negative gradient; 2) with a constant velocity; 3) with a positive gradient. The rectangles indicate the limits on properties of the section (velocities and depths). The shaded areas are the solutions which were obtained.

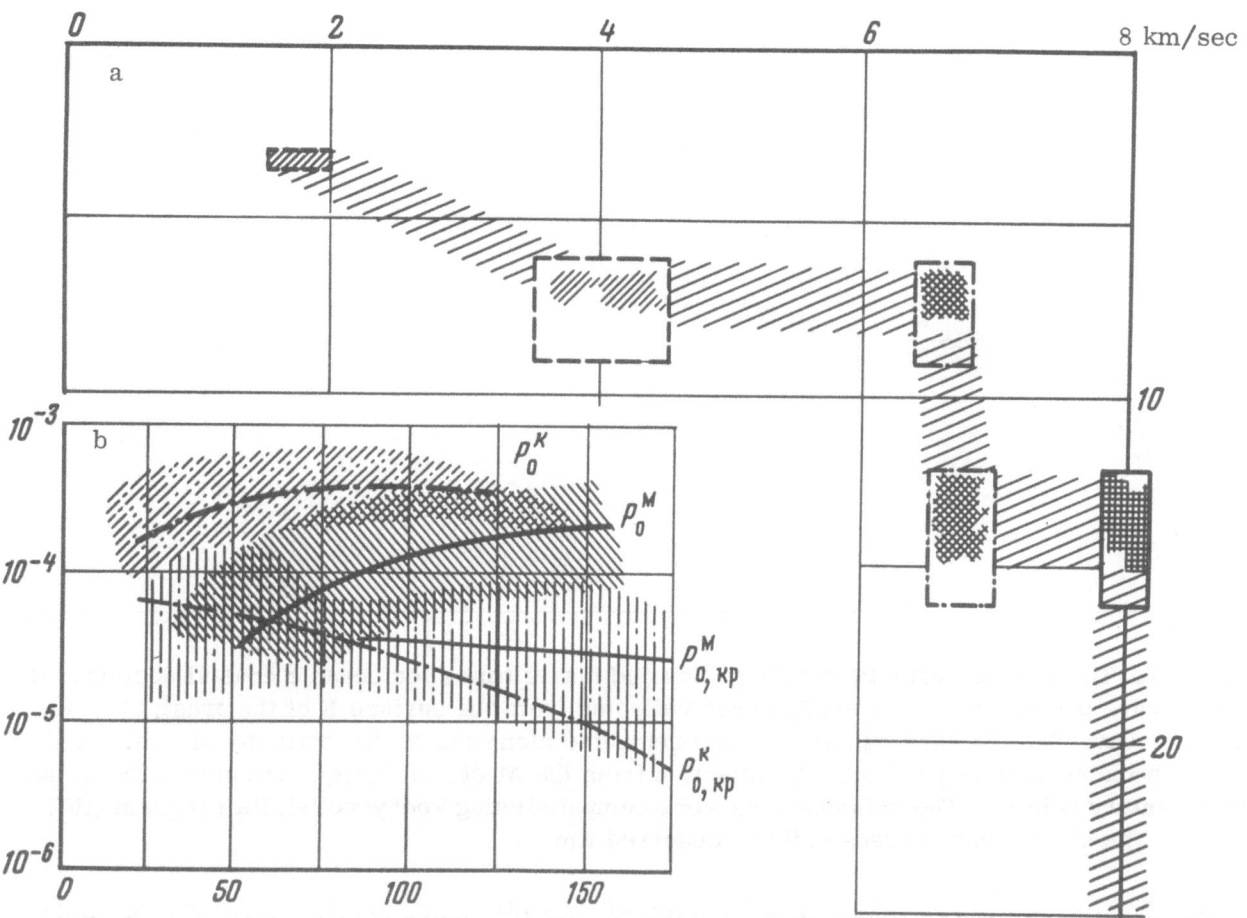

Fig. 5.6. Range of solutions obtained for velocities and amplitudes: a) rectangles (uncertainty specified in data). The shaded areas are the variations in velocity for all models; b) distribution areas for computed wave amplitudes. The curves indicate the amplitudes for single and multiple refracted waves.

and in the mantle (thirty-seven in all) and profiles with a negative gradient (eleven cases). Examples of each type are indicated on Fig. 5.5. Amplitude curves were computed for the P^{sed}, P_0^K, and P_0^M waves, as well as multiple waves $P_{0(KP)}^K$ and $P_{0(KP)}^M$ which have travel times close to those for the primary waves for all sections. The time difference between single and multiple waves was less than the duration of the observed waves, which was about 0.5 sec.

Figure 5.6a shows the area over which all of the velocity profiles so obtained were distributed. It is characteristic that this area is narrower than the range of uncertainty specified for the original data.

The distributions of amplitudes for the various waves are indicated in Fig. 5.6b. The solid curves show the typical forms.

In selecting the most probable sections, we compared the computed and observed travel times and amplitudes. In so doing, it was discovered that curves with negative or large (more than 0.3 km/sec) gradients in the crust and mantle failed to correspond to the observed ranges for occurrence of the P_0^K and P_0^M waves. Some profiles gave very high apparent velocities for the P_0^M wave (more than 9 km/sec), which are not actually observed in the region. For some

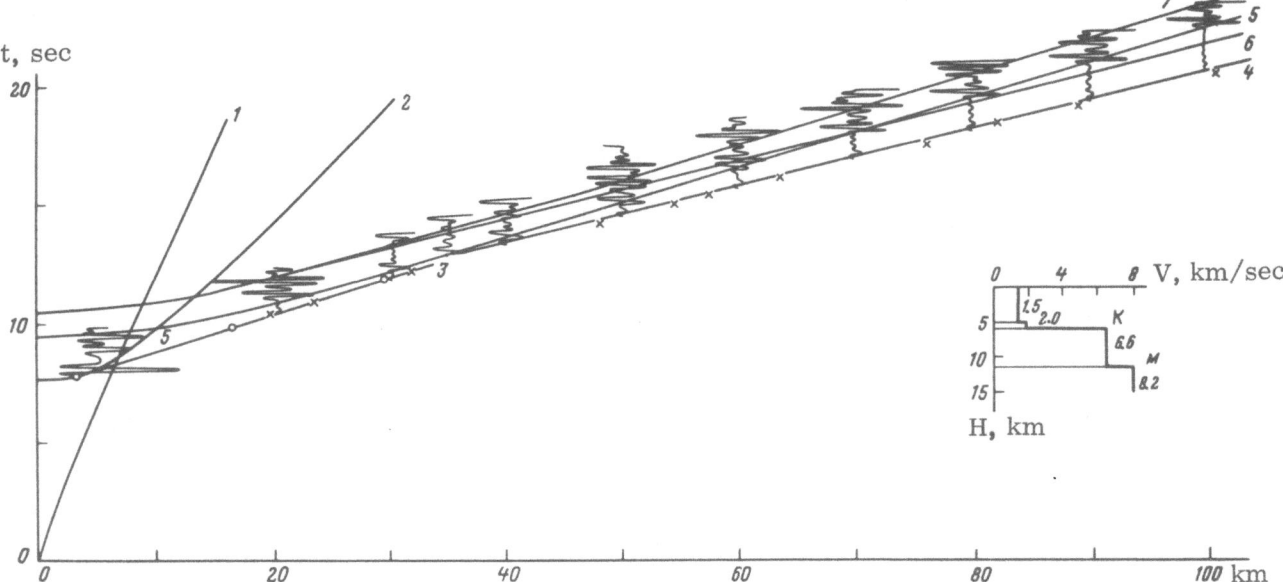

Fig. 5.7. Theoretical synthetic seismograms and travel times for oceanic crustal sections: 1) sound wave propagating in water; 2) waves reflected from the surface K of the crust; 3) wave P_{0H}^K; 4) wave P_{0H}^M; 5) wave $P_{0\,refl}^M$; 6) waves refracted along the M-discontinuity and reflected twice in the sedimentary column; 7) reflection from the M-discontinuity, reflected twice in the sedimentary column. The seismograms were computed using Pod"yapol'skii's program (IFZ, 1964). The circles and crosses indicate observed times.

of the profiles, the ratios of amplitudes for the P_0^K and P_0^M waves did not agree with observed ratios. Based on these data, only eleven profiles were selected from the original forty-eight. These cases corresponded to a positive gradient: no more than 0.03 km/sec/km in the crust, and no more than 0.02 km/sec/km in the mantle.

An evaluation of the possible variations in the sections obtained by calculation is of interest. Such an evaluation may be summarized by the following figures. With a change in the velocity of the upper part of the sedimentary section from 1.5 to 1.8 km/sec and of the lower part from 3.5 to 4.5 km/sec, the depth to the surface of the crystalline part of the crust varies over the range from 6.5 to 8.0 km. With the same changes in properties in the sedimentary section and a change in the velocity of the upper part of the crust from 6.3 to 6.8 km/sec and from 6.5 to 7.0 km/sec in the lower part, the depth to the M-discontinuity changes from 12.2 to 14.8 km.

Thus, the maximum possible error in the deep seismic sounding section attributable to inaccuracies in velocity and the depths to intermediate boundaries amounts to 1.5 km for the surface of the crystalline part of the basement and 2.5 km for the M-discontinuity. If we consider a system of travel-time curves rather than a single travel-time curves, it is possible to narrow the range of velocities and depths and to reduce the band width Δt correspondingly. This also corresponds to the evaluation which has been published previously in [102] for construction of structural maps for deep boundaries.

A comparison of computed and observed amplitude curves supports the hypothesis that the P_0^K and P_0^M waves are usually geometric refractions in areas with a near-oceanic crust, propagating along a medium with a weak velocity gradient. It is required that in such a case the multiple refracted waves will be nearly an order of magnitude weaker than the singly refracted waves. This means that they may be evaluated without being concerned with the contribution from multiple refractions.

In order to improve the model of the oceanic crust, in addition to the solution of the inverse problem, a program written by G. S. Pod"yapol'skii for computing synthetic seismograms was used; that is, the entire seismic wave field was computed at a specified point for a time interval Δt, with interference between all waves being considered. The program was developed for a four-layered medium. The calculations were performed for a uniformly layered section, as shown on Fig. 5.7. The time interval was $\Delta t = 3$ sec. This was found to be adequate for studying the first refracted waves P_0^K and P_0^M, the $P_{0\,\text{refl}}^M$ wave reflected at a less-than-critical angle, and partial multiples of these waves with intermediate reflections in the sedimentary section. The parameters describing the section were assigned by Yu. V. Tulina on the basis of experimental data, and the computations were carried out by E. N. Zaitseva under the direction of the author of the program.

The theoretical travel times for the first arrivals correspond to observed values. However, there are significant differences between the general forms of the computed and observed seismic wave fields. The principal elements in the computed field are the $P_{0\,\text{refl}}^M$ event and interference combinations between it and multiple waves. In the case of the observed field, the $P_{0\,\text{refl}}^M$ wave is not found and the P_0^M wave is usually dominant on the records (see Fig. 5.4d). The P_0^K waves are frequently attenuated more strongly than would be expected from theory, and some times, as is apparent from Fig. 5.2, it (the P* event) is intense and correlation of the event ends abruptly.

The disagreement between the observed field and the theory for a uniformly layered medium confirms the supposition that, despite the fact that in many cases a uniformly layered model of the medium provides satisfactory travel times for the principal components of the field, the amplitudes are not correct.

The observed properties of the records for the principal wave types agree best with the idea of a nonuniformly layered model, even though, as has been indicated in the analysis in Table 19, not all of the many features of the complete seismic wave field for the various types of crust may be explained even so. The analysis of changes in characteristics of the seismic wave field even for a crust of a single type is very difficult.

Thus, investigation of seismic wave fields from different areas of the world has indicated that many variations in the primary field elements are diagnostic for determining the structure of the earth's crust. These properties of the seismic wave field provide the possibility for developing objective criteria for categorizing the crust on the basis of the properties of the medium reflected in the entire wave propagation picture, rather than only from the geometric properties of the section, for which the field may be generalized in a variety of ways.

However, the use of all the properties of the seismic wave field requires development of theory and methods for analysis which will permit us to recognize the elements of the field most characteristic of the principal features of crustal structure — its layering and its blockiness.

§ 2. Regional Classification by Seismic Wave Fields

The correspondence between the characteristics of seismic wave fields and the type of crust is basic to regional classification using deep seismic sounding data from long profiles. Cases in which a uniform density of observation is used are most satisfactory.

Classification of areas according to qualitative and some quantitative characteristics of the records is either the first step in interpretation, done before the waves are correlated, while the nature of the structure in the medium is still unclear and it is necessary to provide a basis for the method used in further analysis of the data, or as a later step, when the travel-time curves and a section have been constructed and it is necessary to supplement them with a qualitative summary in order to obtain a more complete geological and geophysical interpretation.

Such a stepwise approach to the analysis of a seismic wave field has been used widely for marine deep seismic sounding surveys in the Caspian Sea [3] and in the transition zone between Asia and the Pacific Ocean [163, Chapter 2]. Elements of regional classification based on the properties of the field have been used in summarizing deep seismic sounding data from Central Asia [94, 42], Kazakhstan [134], and other regions of the USSR. Recently, a similar approach has been worked out by N. N. Puzyrev for the point sounding method [137-139].

The basis for qualitative analysis of records and its relationship with the geological conditions was laid out by G. A. Gamburtsev as early as 1943 in the study of reef masses in Ishimbai region [49]. Most of the developments along these lines have resulted from development of methods for seismic mapping in vertically structured media (Berzon et al. [23, 24]).

In deep seismic sounding, the regional classification of areas according to the type of data has been done on the basis of an analysis of the following characteristics of the seismic wave field.

1. Record Forms. Complex or simple, long or short, intense or weak, presence or absence of dominant deep-traveling waves among the first arrivals or in the later parts of the records; nature of secondary waves (multiples, shear waves of the type PS_0^KP), properties of the intervals between wave arrivals; short or long, and the character of the coherent and incoherent vibrations in these intervals are considered.

With continuous profiling, record forms differ mainly with respect to the range over which individual waves can be correlated and the amplitude expression of wave groups against the general background of other oscillations.

2. Travel-Time Characteristics of the First Arrivals. a) The travel-time curves for the first arrivals are compared — the actual travel times to various distances from the shot point, the number of segments to the curves, the ratios of apparent velocities for various segments of reversed systems; and b) maps or sections are contoured in terms of equal travel times, $t(R)$. The travel-time cross sections are constructed using the distance R along a traverse as the abscissa, and the travel time t for the first arrival to a specified distance from the shot point as the ordinate. Times are plotted in seconds at distances $R/2$ from the shot point (for marine data, at the recording station).

Contours of time $t(R)$ are plotted on a map in terms of the time t required for the first wave to travel to a distance R with the wave being taken as some diagnostic event, such as the return from the M-discontinuity as the first arrival. In this case the contour map will reflect grossly the relief of this surface if it is assumed that the velocity is constant in the overlying medium.

Such a map is essentially a summary of the travel-time curves for a fixed distance R and serves as an excellent supplement to them.

Time contour maps may also be constructed for the complete system of travel-time curves using the values of t_0 for each segment. The values for t_0 are plotted on the map at the locations of the recording stations. In the case of deep seismic sounding, such a map is usually drawn for the surface of the crystalline part of a crust. c) Graphs for the relationship between average velocity $\overline{V}$ and depth are plotted, or cross sections with contours of $\overline{V}$ as a function of H are prepared. The graphs for $\overline{V}$ vs H are prepared using the travel-time curves for the first arrivals.

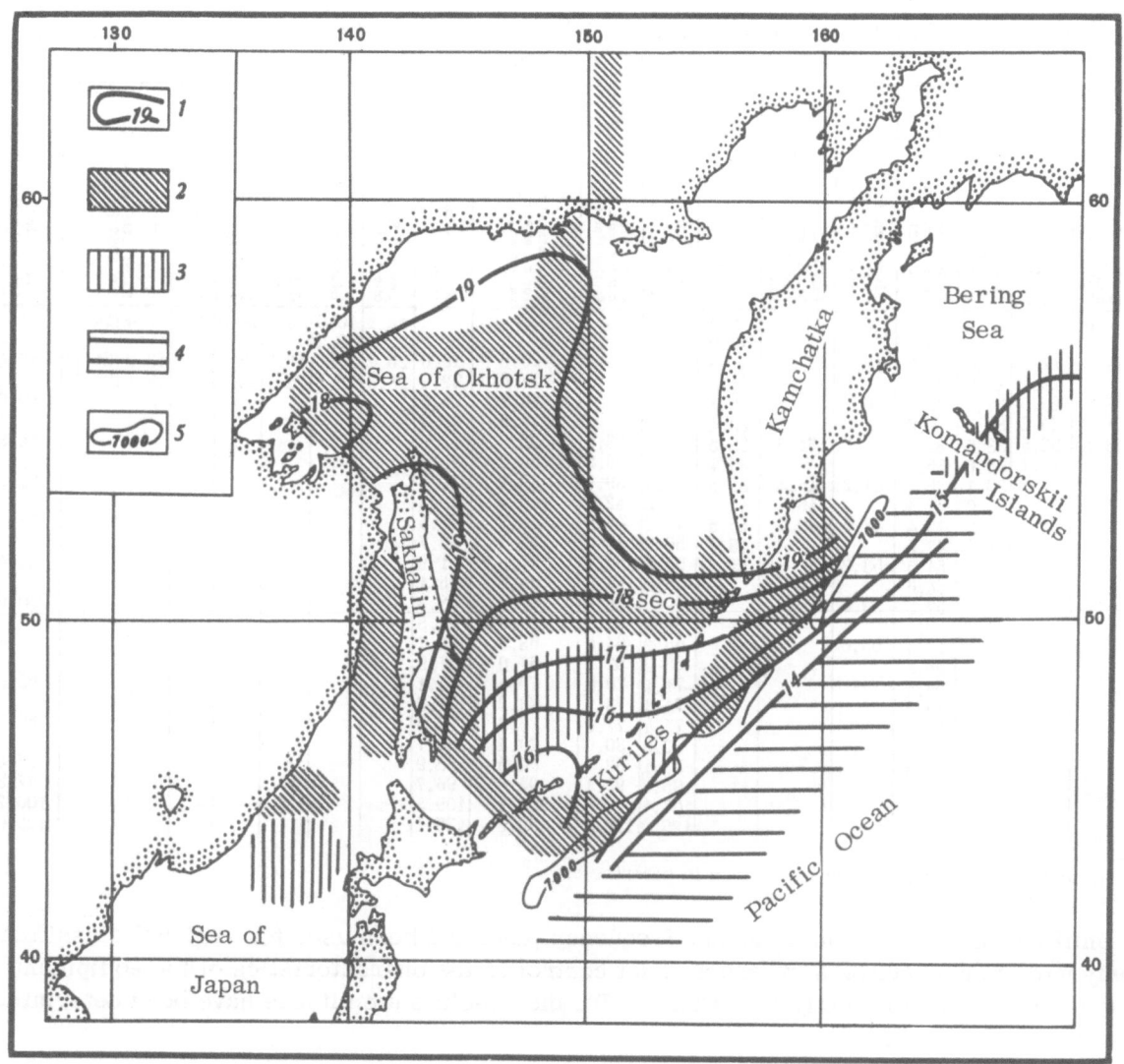

Fig. 5.8. Regional classification of a transition zone on the basis of a preliminary analysis of record types and travel times for the first arrivals (Fig. 5.3): 1) travel times for the P_0^M wave to a distance of 100 km; 2) area with a continental type of seismic wave field; 3) transitional type; 4) oceanic type; 5) 7000 m ocean depth contour.

Such graphs supplement the time contour maps and comparison of the two permits recognition of the increases in travel times on the contour maps related to increase in thickness of each of the layers. The contours of $\overline{V}$ as a function of H on the cross sections are prepared from the graphs of $\overline{V}$ as a function of H. Methods for their construction have been described in [163, Chapter 2].

3. Amplitude Characteristics. Plots of the amplitude A and frequency f as a function of distance R are analyzed. The amplitude plots A(R) for the first arrivals permit a more reliable determination of the number of travel-time curve segments and a determination of the general character of energy attenuation in the medium. Areas with anomalous increases or decreases in intensity are separated from the rest of the records. Frequently, these plots provide some idea of the effect of the sedimentary section and the resonant properties of the deep boundaries on the seismic wave field, f (R) (see Chapter II).

TABLE 21. Travel Times (in Seconds) for the First Arrivals in Deep Seismic Sounding and Seismological Profiling

Distance, km	Baltic Shield, Kem-Ukhta profile [113]. Island shot point	Russian Platform, summary travel-time curves [131]	Caucasus, Stepnoe to Leninakan. Shot point at Stepnoe	Turkmenia, Ashkhabad to Aral Sea [108, 154]. Shot point 117	Tien-Shan, Issyk-Kul' to Balkhash [50]. Shot point at Issyk-Kul'	Pamir-Alai zone, Daraut to Kurgan to Irkeshtam [96]. Shot point at Kabud-Khauz	Kazakhstan, Temir-Tau to Petropavlovsk [8].	Seismological travel-time curves from the Pamir to the Lena River [127]	Seismological travel-time curves from the Lena River to the Pamir [127]	Theoretical travel times for an average continental crust [223]	Jeffries-Bullen travel time	Caspian Sea, Zhila profile, one station [3]	Black Sea [124]	Okhotsk Basin, profile 6M, 1957 163	Bering Sea, profile 8M 1958 [163]	Sea of Japan, profile 26, 1964	Pacific Ocean, profile 1-0, 1957 [163*]	Theoretical travel times for an average oceanic crust [223]
50	9.0		13.4	10.9			8.6					14.1	10.0	9.2	10.0	9.1	7.4 (14.1)	7.2
75				15.5			12.9					22.2	13.2					
100	17.0	17.5	20.0	19.5	18.6		17.8			20.6	18.0	26.2	20.1	16.6	16.3	15.7	13.2 (21.1)	13.4
125		22.1	23.7	23.0			21.3			23.6	22.4	28.8	22.1	22.7	22.5	22.4	19.1 (26.7)	
150	24.5	26.0	27.5	26.5	26.5		25.3			26.7	26.9							19.6
175	28.5	29.2	31.0	29.8			30.4			29.8	31.4							
200	31.5	32.5	33.5	33.0	34.3	34.5	33.5		33.9	32.9	32.5							25.8
225		35.7	35.6	36.2			36.7		36.9	36.0	35.9							32.0
250			39.1	39.4	41.2	43.4	39.5	41.2	39.9	39.1	39.0							
275			43.0	42.4		46.6	42.7	44.2	42.8	42.2	42.3							38.2
300			45.8	45.4	47.6	49.2	45.6	47.2	45.8	45.3	45.5							
325			48.5	47.5			48.4	50.2	49.0	48.4	48.7							44.4
350				51.5	53.2		51.2	53.2	52.1	51.5	51.9							
375				54.6				56.2	55.2	54.6	55.1							50.6
400				57.5	59.0			59.2	58.4	57.7	58.3							
425				60.2				62.2	61.5	60.8	61.5							56.8
450				63.4	65.0			65.2	64.6	63.9	64.7							
475				66.8				68.2	67.7	67.0	67.9							63.0
500				69.8				71.1	70.8	70.1	71.1							
525				73.5				74.1	73.9	73.2	74.3							69.2
550				76.9				77.1	77.0	76.3	77.5							
575				79.4				80.0	80.0	79.4	80.7							75.4
600				82.4				83.1	83.1	82.5	83.9							
700								95.0	95.1	94.6	96.7							87.8
800								107.1	107.6	106.7	109.5							100.2
1000								125.0	126.3	130.9	135.1							125.0

*Times without correction for sea depth are given in parentheses.

Until recently, such amplitude and frequency plots had been used for regional classification only with marine surveys, where a strict control of the characteristics of the equipment was maintained. More recently (see Chapter II), the absolute amplitudes have been determined with land surveys [1, 8].

Usually, regional classification according to the type of data is not difficult when areas which differ markedly in crustal structure are being studied, such as contact zones between areas of platform and geosynclinal types. The procedure becomes much more complicated within an area with a single type of crustal structure, as for example, a platform area or a shield area.

Areas which exhibit the various types of behavior for seismic wave fields across a transitional zone are shown in Fig. 5.8. This classification is from [163, Chapter 2]. It was compiled in 1959 when interpretation of the data was in an early stage and was supplemented by new data in 1963 and 1964.[†]

Three principal types of behavior were recognized on the basis of an analysis of the seismic records and travel times for the first arrivals. A contour map of t(R) for R = 100 km was prepared from the travel-time curves. In most of the area, the first arrival at this distance was the P_0^M event. Despite the fact that this approach is overly simplified, the results reflected the general character of crustal structure in the transitional zone rather well. This approach proved to be quite useful as a first step in interpreting deep seismic sounding data obtained over a broad transitional zone.

[†] Data from the Pacific Ocean Deep Seismic Sounding of the Institute of Physics of the Earth and the Sukhalin Joint Institute, 1963-1964.

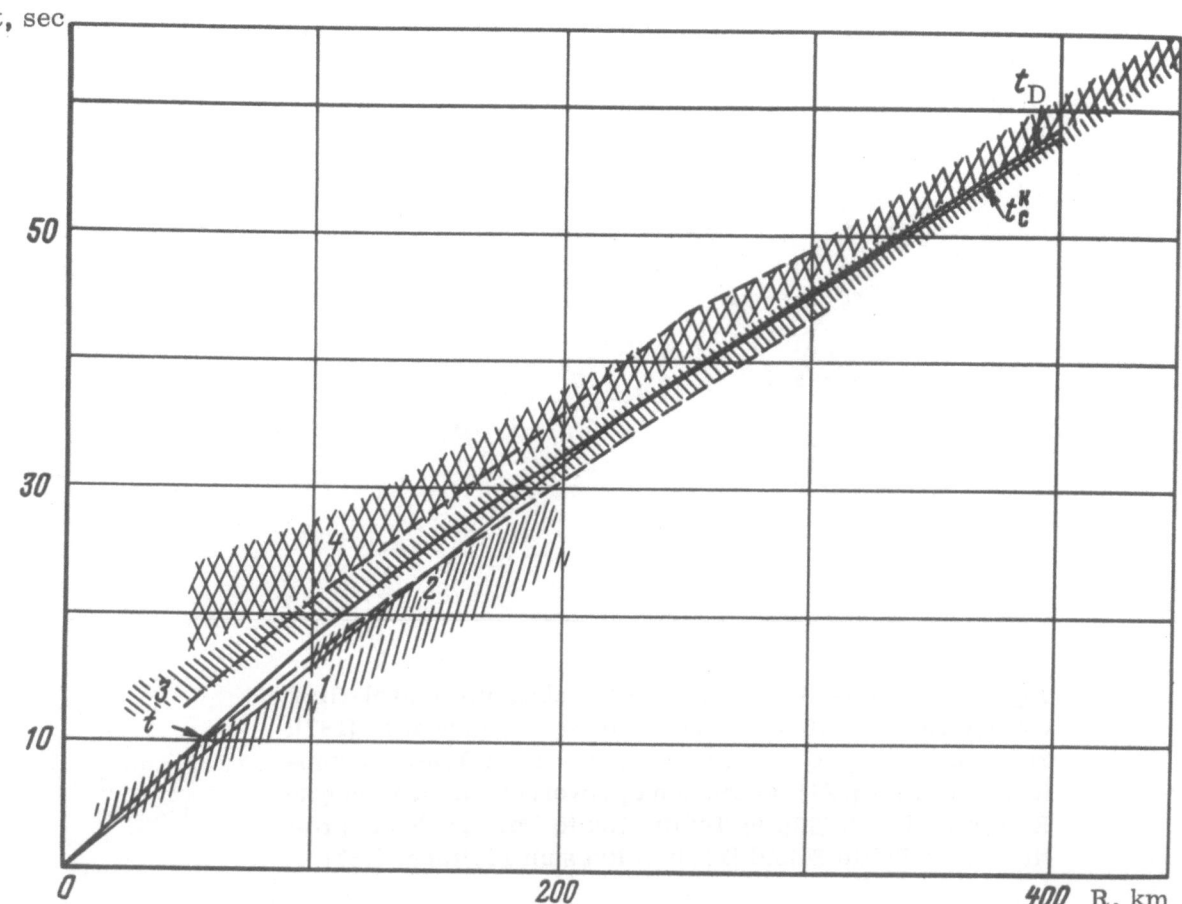

Fig. 5.9. Travel times for the P_0^M and the P_{refl}^M events for oceanic and continental structure: 1) range of travel times for an oceanic structure (corrected for sea depth); 2) the same, for fringing seas; 3) the same, for high-angle reflected waves $P_{0\,refl}^M$ and refracted waves P_0^M for shielded and platform areas; 4) the same, for continuous regions and deep basins with more than 10 km of sediments (southern part of the Caspian Sea, Predkopetdag Basin); t_D are the Jeffries−Bullen travel times for first arrivals in an average continental crust; the dashed lines indicate first arrivals in Alpine travel-time curves [202].

Classification on the basis of the type of data obtained is now done with all deep seismic sounding efforts. However, the approach is most effective when it is used as a first step in interpretation in studying a broad region, where it may be necessary to obtain a rough idea about the relationships of major changes in the character of the data in going from the tectonic zone to another (Kosminskaya [103; 163, Chapter 2]).

§ 3. Summary Travel−Time Curves for Various Tectonic
Zones in a Continental Crust

So far, we have considered the characteristics of seismic wave fields for various types of crustal structure and have illustrated these with examples from a transitional zone. There is an equal amount of concern with the analysis of seismic wave fields within tectonic zones of a single type. Evaluation of deep sounding data from this point of view has only just begun. Data are available from within the USSR which permit an examination of the properties of first-arrival travel times for several types of tectonic zones and comparison of these with the

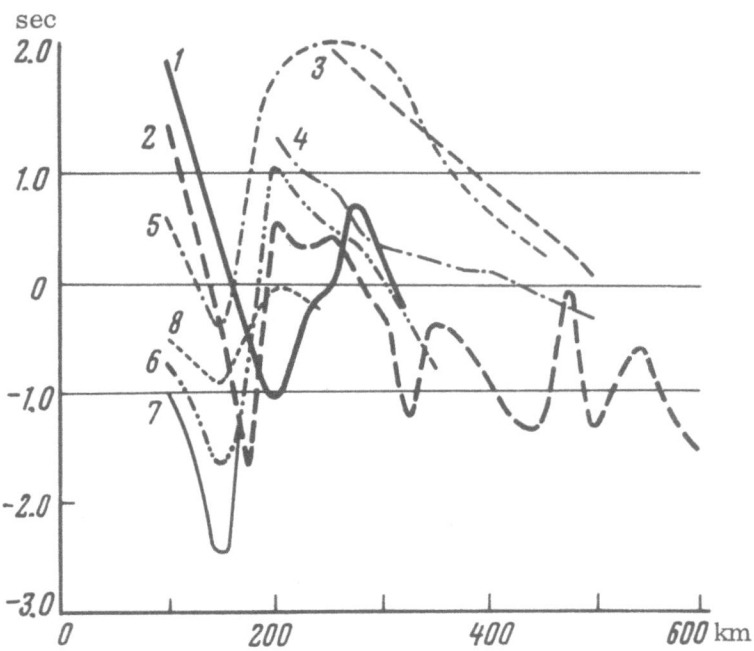

Fig. 5.10. Time differences between observed travel-time curves and the Jeffries—Bullen curves: 1) Caucasus [187]; 2) Predkopetdag Basin [42]; 3) Pamir—Lena River seismological profile [127]; 4) the same, reversed [127]; 5) Issyk-Kul' to Balkhash [50]; 6) Temir-Tau to Petropavlovsk profile [8]; 7) Baltic Shield [114]; 8) Russian Platform [131].

Jeffries—Bullen statistical travel times t_D and the t_c^K travel times. We have computed the t_c^K travel times for a statistically averaged continental crust using data summaries [223].

Table 21 is a list of travel times for first arrivals for the Baltic Shield, Russian Platform, Kazakhstan, Turkmenia, and mountainous regions of the Caucasus, Tien-Shan, and the Pamir-Alai. Seismological travel times compiled by I. L. Nersesov and T. G. Rautian [127] for the Pamir—Lena River profile and the reversed Lena River—Pamir profile are included (see Fig. 5.12). Differences in travel times for various regions are readily apparent in the table.

Travel-time curves for various zones collected into groups are shown in Fig. 5.9. Four groups are recognized: oceanic (1), near oceanic (2) and two continental groups — for shields, platforms, and the Kazakhstan Basin (3), and for mountainous areas and deep basins (4).

The well-known Jeffries—Bullen travel-time curves for an average continental crust (crustal thickness of 33 km, $H_1 = 15$ km, $V_1 = 5.5$ km/sec, $H_2 = 18$ km, $V_2 = 6.5$ km/sec and $V_3 = 7.9$ km/sec) and t_c^K travel-time curves (crustal thickness of 40 km, average velocity in the crust of 6.25 km/sec, and 8.1 km/sec at the M-discontinuity) are shown in Fig. 5.9. A comparison of these travel-time curves indicates that they differ significantly at distances from 50 to 200 km, are nearly coincident at distances of 200-300 km, and diverge again at distances greater than 300 km.

We will examine Table 21 in detail, as well as Figs. 5.10 and 5.11, which are plots of time differences between observed travel-time curves, and the average curves t_c^K and t_D.

Among the continental travel-time curves, the largest time differences are found for the distance range 0-150 km, where the S and K waves comprise the first arrival.

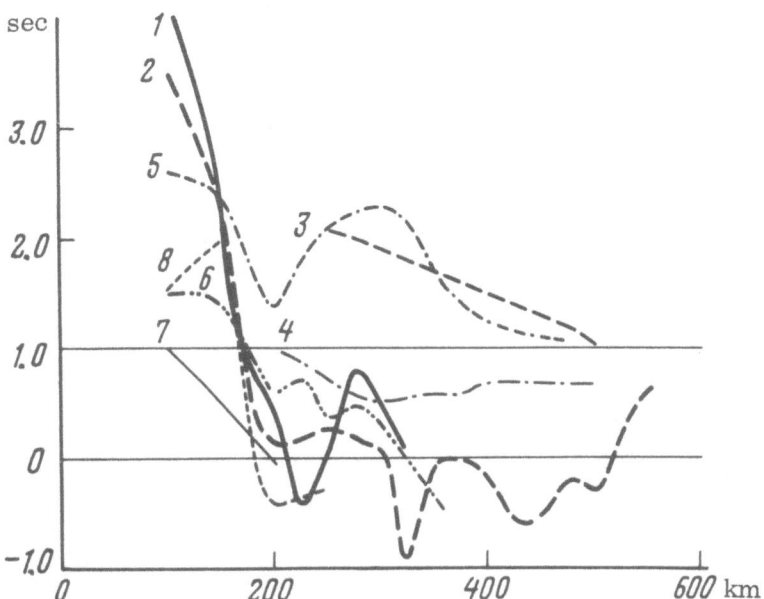

Fig. 5.11. Time differences between observed travel-time curves and the t_c^K curves (see Fig. 5.9). The numbers are defined for Fig. 5.10.

In regions with a thick sequence of bedded sediments and a correspondingly large travel time for the five events and long interval over which the first arrival is an S event, the travel-time curves show the greatest travel times (South Caspian Depression), and in areas of basement outcrop and shield areas (Baltic Shield and Kazakhstan), the travel times are least. The difference in travel times in the two cases amounts to as much as 10 sec over the distance range 50-100 km (Table 21).

The travel times for the Caucasus, Predkopetdag, Black Sea, and Fergana Basins occupy an intermediate position in the continental group.

The differences between travel-time curves in the continental group at distances of 150-300 km are small and do not exceed 3-5 sec. Such differences are characterized by the differences in travel times for the mountainous and platform areas in Central Asia. These are in close agreement with the seismological travel times along the profile from the Pamir to the Lena River (Fig. 5.12).

In the case of oceanic and near-oceanic travel times, data on travel times for the first arrivals are limited to distances less than 200 km. The difference between oceanic and continental travel times practically disappears if no correction is made for sea depth, but the actual travel times are used.

A marked change in slope of the travel-time curves over the range of distances 100-200 km, that is, in the region where the P^M event emerges as the first arrival, is also characteristic of all of the data, and the data behave with greater uniformity at greater distances.

The form of the curves at distances up to 150 km is explained by the influence of the sedimentary section, which causes the increase in travel times for the first arrivals over this range. This effect becomes gradually less at greater distances, and time differences decrease correspondingly.

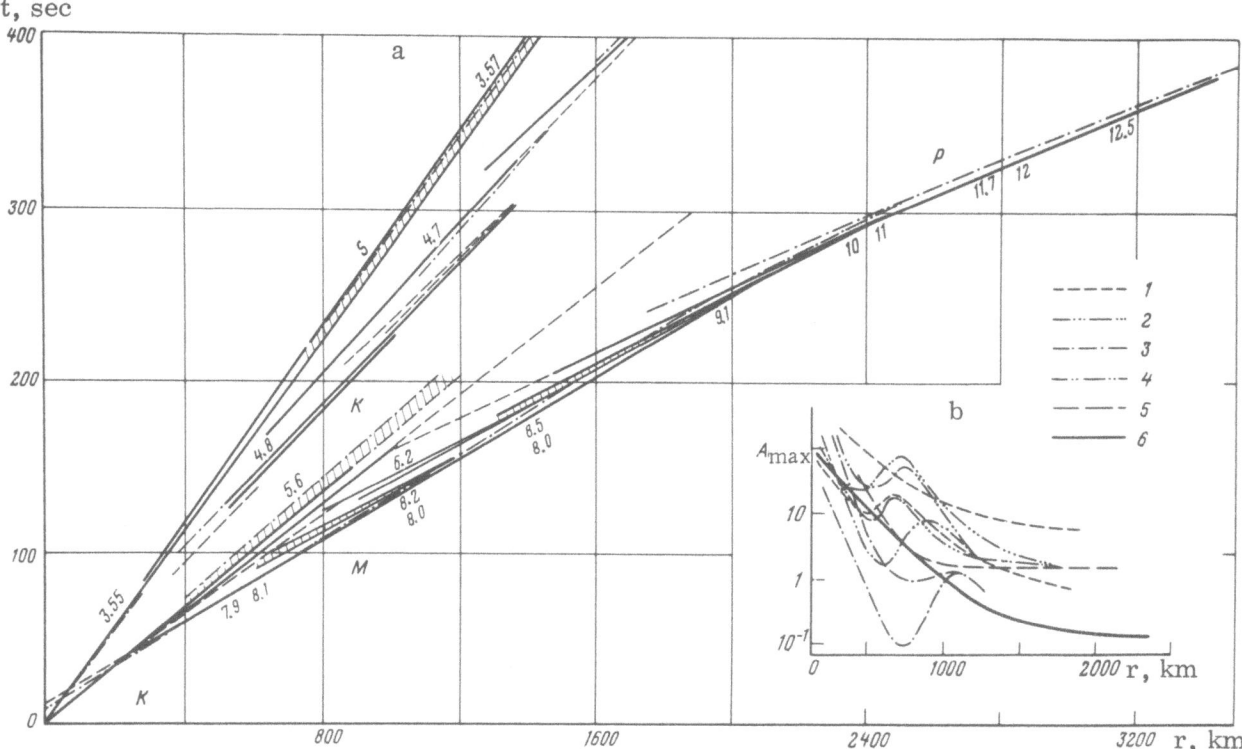

Fig. 5.12. Travel times (a) and amplitudes (b) for K and M events, based on seismological data. Profile from the Pamir to the Lena River (Nersesov and Rautian [127]): 1) Central Asia; 2) Dzhungariya; 3 and 4) Altai, Sayan; 5) East Sayan; 6) near Baikal.

Analyzing these curves, we may conclude that the maximum deviation between the observed travel times for distances up to 200 km amounts to ±2 sec. Over distances from 200 to 500 km, these deviations are largely positive, amounting to +2 sec in mountainous areas, and less than 2 sec in other areas.

Nearly all of the departures from the t_c^K travel-time curves are positive. They amount to 4 sec over the range from 100-200 km. At larger distances, they amount to 2 sec in mountainous areas, but are less than 1 sec in platform areas. Hence, it follows that the t_c^K travel-time curve is closer to the observed data in the later case than the Jeffries—Bullen t_D curve.

Plots of time differences $\Delta t_D = t_H - t_D$ and $\Delta t_c = t_H - t_c$ were prepared from the reversed seismological travel-time curves along the profile between the Pamir and the Lena River in order to compare the observed and theoretical travel times at larger distances in the range over which layers in the upper mantle exert an influence (Fig. 5.13). It is apparent from the plots that the t_c^K is closer to the observed travel times than the Jeffries—Bullen curve. Because this travel-time curve is a better approximation to observed travel times in shields and platform areas, which comprise a large portion of the continents, we propose the use of the t_c^K travel-time curve for improving the accuracy of seismological data from near earthquakes.

The summary of deep seismic sounding travel-time curves in Figs. 5.10-5.12 makes it possible to determine which type of crustal structure is being dealt with from a very early analysis of travel times for the first arrivals when work is started in a new area. In the case of seismological investigations, they make it possible to evaluate errors related to the influence of regional-scale inhomogeneities in the earth's crust.

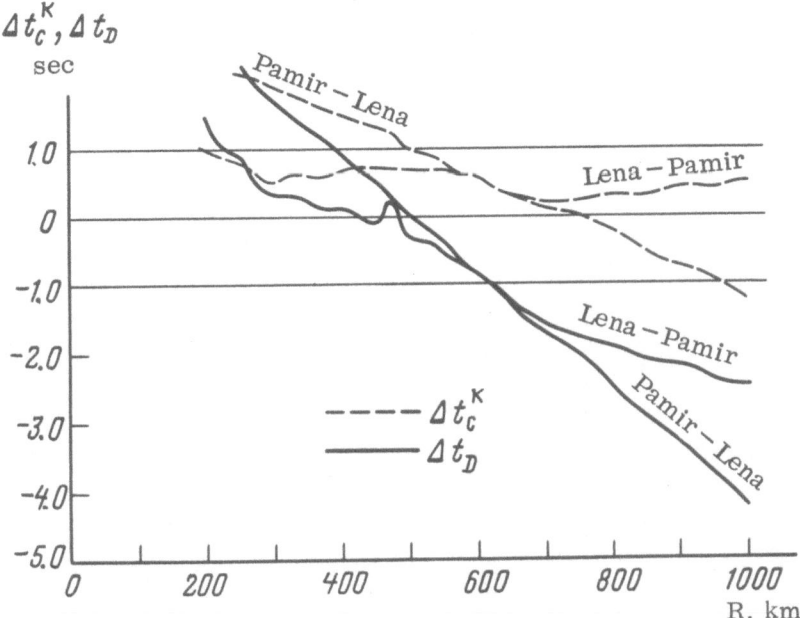

Fig. 5.13. Time differences for seismological travel times observed along profiles from the Lena River to the Pamirs, and the reverse (Nersesov and Rautian [127]), and the t_D and t_c^K travel-time curves.

A further generalization of deep seismic sounding data is of interest, particularly with respect to first arrival travel times for crustal and sedimentary waves over the range from 0 to 150 km which characterize the relationship of the sedimentary section and the upper part of the crystalline crust, and for the range from 200 to 500 km, where the first identifiable M waves are related to the upper mantle at depths of 100-120 km. Data on deep seismic sounding travel times may supplement and improve the reliability of the results from seismological time-distance studies from the range of distances from 500 to 1000 km, which characterize the mantle at depths of 100 to 200 km, the study of which is particularly interesting in relation to investigations of the asthenospheric layer with low velocity. In such a plan, it would be highly desirable to carry out detailed seismological studies over ancient sheets and platforms, where recently it has been suggested on the basis of low heat flow values that an area of decrease in velocity of longitudinal waves with depth in the mantle is absent [201].

CHAPTER VI

TYPES AND STRUCTURES OF THE CRUST

In an earlier chapter, it was indicated that as the frequencies of the seismic waves used in probing the crust are raised, the behavior of the observed seismic wave field becomes more complicated. Despite this, there are two basic features of the crust and upper mantle at all frequencies: layering and blockiness.

In this chapter, we will examine the basic properties of the crust derived not only from sections but also from more objective characteristics which are reflected in the principal features of the seismic wave field.

In order that we may obtain a good idea about the degree of approximation in the earlier seismic data, we will examine briefly some new concepts about the crust and upper mantle. Further insight into the types and structures of the crust will evolve from these newer concepts. In particular, we may gain insight into the nature of layering in the crust.

§ 1. New Ideas on the Structure of the Earth's
Crust and Upper Mantle

The amount of work done in studying the crust under the continents and oceans in the last three to five years is two to three times greater than that done in the preceding decade [38, 103-105, 146, 147, 235, 238]. In 1966, 150 deep seismic soundings were completed in the USSR, while in 1960, the number had been about 40. The total length of profiling completed in the USSR (to the end of 1966) is about 60,000 km. More than 1200 crustal sections over the entire world have been completed.

Information on the deep structure of the crust and mantle is not limited to that derived from seismic data. Such data are frequently combined with other types of geophysical data and information about the structure of the earth. However, deep seismic sounding is the primary and most satisfactory source of data.

The velocity profile as a function of depth and horizontal changes in this profile are of fundamental importance in understanding crustal structure. The profile has been obtained in the USSR primarily from detailed deep seismic sounding surveys along continuous profiles on land and during marine operations in the Pacific Ocean, and its transitional zone, as well as on the inland seas — the Black Sea and the Caspian Sea.

Development of the theory for interpreting seismic wave behavior and methods for analyzing this behavior, a subject which was covered in Chapters IV and V, has resulted in new interpretations of the velocity profile in the crust and upper mantle. New concepts about the structure of the crust that are essentially different than earlier concepts have been developed on the basis of the new data. This evolution of ideas may be summarized briefly as follows:

1. The Continental Crust May Be Thicker and the Velocity Higher Than Previously Supposed. Based on new data, the average thickness of the crust under the continents is 40 km, and the thickness of the crust under shield and platform areas is 35 to 40 km, while according to the Jeffries—Bullen data, the crustal thickness was taken to be 33 km, and under shield and platform areas, 25-30 km.

New data give evidence on an increase in velocity in the continental crust from about 6.0 km/sec at the surface to 7.2-7.6 km/sec at the base of the crust, with the layer having a velocity of 6.0 km/sec being thickest. The layer with a velocity of 6.0 km/sec never exceeds one-third the total crustal thickness in shield and platform areas [114, 131, 160]. It was supposed earlier, and this followed logically from interpretation of the earlier less precise seismic data, that in the limit, half of the crust consists of a layer with a velocity of 5.5-6.0 km/sec.

Essentially, the ideas about the role of the sedimentary layer in the crust have changed. It is recognized to be even more irregular, and thicker on the average. Regions are known where the sedimentary layer comprises 2/3 of the total thickness of the crust, and has a velocity of 3 to 4 km/sec [3, 97, 99, 124].

2. A Shift from One- or Two-Layer Models for the Crust to Multiple Layered Nonuniform Models. Three or four boundaries with velocity increases of about 5.8-6.2, 6.3-6.5, 6.8-7.2, and 7.4-7.7 km/sec are recognized in the continental crust in most regions. It has been shown that velocity layering is a typical phenomenon for the crust in various tectonic zones. Low velocity layers may also be present in the crust, but such layers have not yet been recognized from deep seismic sounding data.

Areas where layering cannot be recognized are anomalous in this general picture [8].

3. Concepts about the M-Discontinuity as a Sharp and Well-Defined Boundary between the Crust and Mantle Have Changed. While it had previously been supposed that the velocity discontinuity at the Moho boundary was 1.5-2.0 km/sec and was characterized by a velocity ratio of 0.8, now it is apparently necessary to assume that under the continents this discontinuity does not exceed 1.0 km/sec, and the velocity ratio is 0.9 or higher. For the oceanic areas, the difference between the velocities in the crust and mantle is larger and amounts to about 1.5 km/sec.

In many cases, a layer with a velocity of 7.4-7.6 km/sec is recognized above the M-discontinuity, and commonly this boundary is clearer seismically than the underlying boundary that is characterized by a velocity of 8.2-8.4 km/sec. In such cases, a question arises in defining the boundary between the crust and mantle as to which boundary should be called the Moho and this had lead to some controversy.

4. Concepts about the Structure of the Upper Mantle Have Changed (this includes the first 20 to 60 km beneath the M-discontinuity). In contrast to its previously supposed uniformity and homogeneity the mantle also appears to be layered, at least in the very topmost portion [153]. Here, the seismic layering of the mantle is similar to that in the crust. In the mantle, only the general level of velocity changes, and the relative differences in velocity are still smaller. The velocity ratios between individual layers in the upper mantle are close to unity, being 0.95-0.99.

A characteristic of new data on the upper mantle is that the rate of increase in velocity with depth which begins immediately beneath the M-discontinuity is greater than was supposed earlier. Under the continents, deep seismic sounding data have indicated a layer with a velocity of 8.6-9.0 km/sec[†] at a depth of about 100 km, and at a depth of about 20 km under the

---

[†] These velocities, as well as those in the crust, may correspond to individual zones with high velocity. Apparently there may be occasional boundaries with higher velocity contrasts, as evidenced by interference effects.

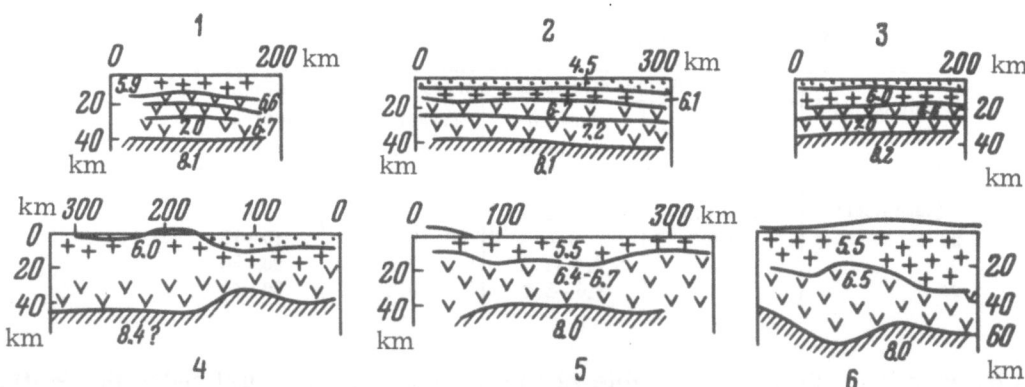

Fig. 6.1. Crustal sections. Shields and platforms: 1) the Baltic Shield (Litvi-nenko et al. [114]); 2) Russian Platform (Pomerantseva et al. [131]); 3) Turk-menian Platform (Krasnopevtseva et al. [42]). Active areas: 4) Caucasus (Yurov [187]); 5) northern Tien-Shan (Gamburtsev et al. [50]); 6) Pamir—Alai region (Kosminskaya et al. [96]).

oceans, whereas according to Gutenberg's curves, such high velocities should be present at a depth of about 400 km, and he recognizes a low-velocity asthenospheric layer in the upper mantle (with a depth amounting to 100-200 km under the continents, and to 60-100 km under the oceans [201]). Deep seismic sounding data on the structure of the upper mantle may be used to clarify its structure at great depths, particularly in the asthenospheric layer, by using sur-face waves in studying it, as well as body waves from earthquakes and large explosions in the shadow zone.

5. Blockiness and Macroinhomogeneities are Present in the Crust and Upper Mantle in Addition to Layering. In most cases, different tectonic zones correspond to different crustal blocks, with differing crustal thicknesses, velocity char-acteristics, and differing relative thicknesses of the sedimentary layer and the crystalline part of the crust [101]. These primary blocks or macro-inhomogeneities of the first order fall into groups on the basis of less obvious characteristics: the nature of the relief on deep boundaries, the degree of sharpness in the transition from one velocity layer to another, and so on [27, 39, 66, 115, 149, 180, etc.].

The boundaries of these blocks are presently recognized from deep seismic sounding as reasonably wide belts in which steeply inclined surfaces with complicated relief are found. Sometimes, these regions are characterized by a distortion of the seismic wave field, rep-resentative of so-called steps or fractures. Within a single block, a deep boundary is either nearly horizontal, or undulates slightly [27]. Specific relationships have been established be-tween the blocks of crust and upper mantle, so that in some we may speak of combined crust and mantle blocks [39, 149; 163, Chapter 12].

We will now present examples of crustal sections obtained with the deep seismic sounding methods in order to illustrate these fundamental concepts.

Examples of a Layered Crust. Crustal sections for the Baltic Shield, the ancient Russian Platform and the young epi-Hyrcenian Turkmenian Platform are shown in Fig. 6.1. All of these profiles were obtained with continuous profiling, and each is characterized by essentially the same detail.

Three boundaries are present in the crust, excluding the upper boundary above which the sedimentary layer is present.

TABLE 22. Typical Velocity Sections

Number of boundary	Boundary velocity, km/sec			Depth interval, km
	Baltic Shield	Russian Platform	Turkmenian Platform	
1 (basement)	5.4—6.2	6—6.3	5.5—6.3	0—5
2	6.6	6.7	6.6	10—15
3	7.0	7.1—7.2	7.0	20—30
M	8.1	8.0—8.1	8.2	30—40

As may be seen from Fig. 6.1 and Table 22 the thickness of the crust under the Baltic Shield and the ancient Russian Platform is greater than the thickness under the young Turkmenian Platform. However, the depth to the intermediate boundaries is the same for all three regions.

The clarity of the velocity characteristics for boundary 2, which is usually called the Conrad Discontinuity, should be noted. However, as was indicated in Chapter IV, with studies made at the frequencies corresponding to seismological data, this boundary may be completely absent, or if it is present on the records as a later arrival at considerable distances from the source, the depth to the boundary is significantly different than that obtained from deep seismic sounding data.† The other intermediate boundaries are not recognized from seismological data.

In considering the sections which have been shown here, the question of the determinacy of the M-discontinuity also arises. According to A. V. Egorkin's data [69], in western Turkmenia, two boundaries are recognized in the lower part of the crust; the velocities are 7.6 and 8.4 km/sec. The first (see Fig. 4.12) is situated at a depth of 35 to 37 km, while the second is at a depth of about 45 km. It is not now clear which of these boundaries should be correlated with the M-discontinuity recognized on sections from other regions.

It is possible that it would be more nearly correct to call the lower boundary the top of the mantle. In this case, the Turkmenian Platform being younger, would be characterized by a greater crustal thickness, which is quite logical considering the data indicating a greater crustal thickness in deep basins and for mountainous geosynclinal areas.

Examples of Mountainous Crust. At present, we have considered deep seismic sounding data from two active regions: 1) Tien-Shan and the Pamir; 2) the Caucasus (Fig. 6.1). These are data obtained with relatively poor detail. It is possible only to compare such general characteristics as the total thickness of the crust and relief of the M-discontinuity.

The complexity of the relief of the M-discontinuity is a diagnostic feature of the crust under mountain systems. The concept of mountain roots is applicable only to major systems of mountain ridges. A consistent set of local uplifts and downwarps of the M-discontinuity, which characterize ridges and valleys in the upper mantle, is observed over an entire active region. As yet, there are not adequate data from deep seismic sounding on the velocity characteristics of the M-discontinuity in active zones, which is a result not only of the paucity of deep seismic sounding data, but also of the operational difficulties involved in obtaining and interpreting data in mountainous areas.

Examples of Sections with a Thin Sedimentary Layer. The crust under the forelands of mountainous areas and intermontane basins, as well as that beneath inland seas with a near-

---

† This is usually related to the erroneous picking of the strong seismological event P* as the head wave from the Conrad, when the actual origin of this wave is considerably more complex (see [6, 220]).

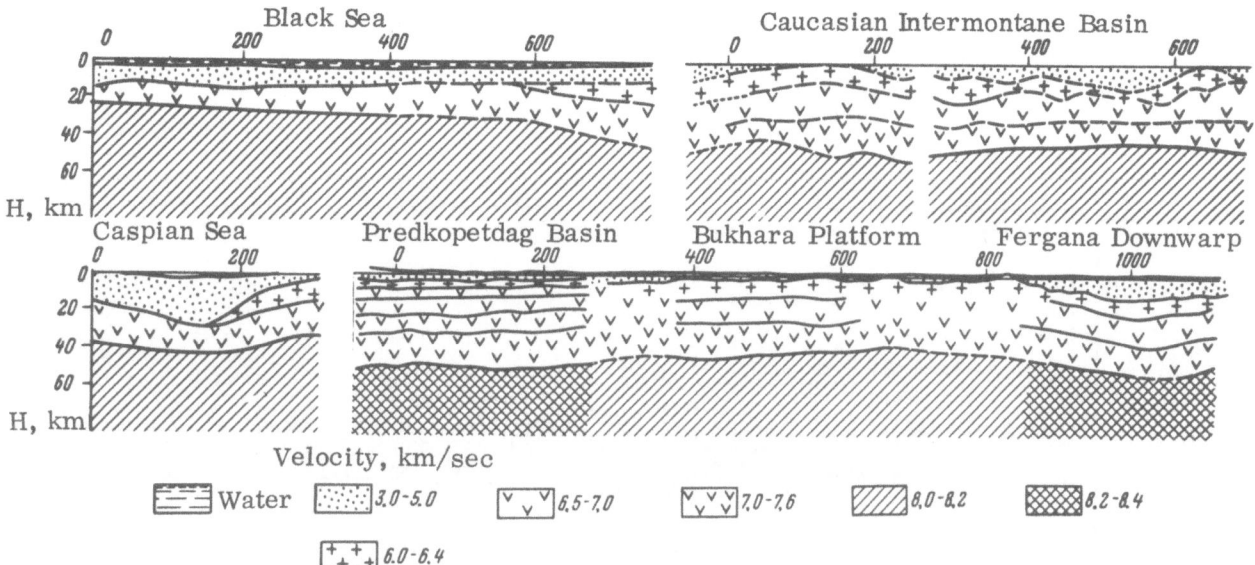

Fig. 6.2. Crustal sections for inland seas and basins [106].

oceanic type of crust, has its own diagnostic features (Fig. 6.2). The main distinction in basins is the thick sedimentary layer, which reaches 10 to 15 km in thickness in mountain forelands and intermontane basins, and more than 20 km in the South Caspian Basin. This layer contributes diagnostic features to the seismic wave field as a result of the presence of a large number of low velocity waves related to boundaries within this layer.

A second characteristic of these sections is the high velocity in the crystalline part of the crust, and in some cases, in the upper part of the mantle as well (the Predkopetdag basin). These regions are characterized by well-defined first arrivals with a velocity of 6.6-7.0 km/sec and a small interval over which arrivals with a velocity of 6 km/sec can be traced (Caucasus, Kurin Depression [64]). In the Black Sea and Caspian basins, waves with a velocity of 6 km/sec are completely absent from the records. Here, the first recorded waves have a velocity of 6.4-6.8 km/sec [124].

Velocity Inhomogeneities at the M-Discontinuity. The preceding examples have illustrated the complexity of the relief of the M-discontinuity. We will now present data on the velocity characteristics for various tectonic zones of the continents and oceans.

Figure 6.3 shows a summary of data on velocities at the M-discontinuity and on its depth. These data indicate that the value for a velocity of 8.1 km/sec, accepted in seismological studies, is a good average for data for continental zones. There are significant departures from this average value toward larger values in basin areas. There is no systematic decrease in the velocity of the upper mantle with depth of the M-discontinuity, as has been indicated to exist by Gutenberg [62]. On the contrary, the greatest velocities along this boundary are related to oceanic areas where the depth is small. The maximum anomaly in velocity is noted in transitional zones, and, in particular, in the vicinity of volcanic island arcs; and moreover, oceanic areas are characterized by high velocities while near-oceanic island areas are characterized by low velocities. Anomalously low velocities (7.6-7.8 km/sec) are noted for the mid-ocean ridges. This is a region of so-called crust—mantle mix.

Small areas (20-30 km) with increased (8.4-8.6 km/sec) or decreased (7.6-7.8 km/sec) velocity have been noted in many areas where deep seismic sounding continuous profiling has been done (the Ukraine, Kazakhstan, the Urals). Possibly the most complex blockiness to the M-discontinuity has been noted in the southern Kuriles.

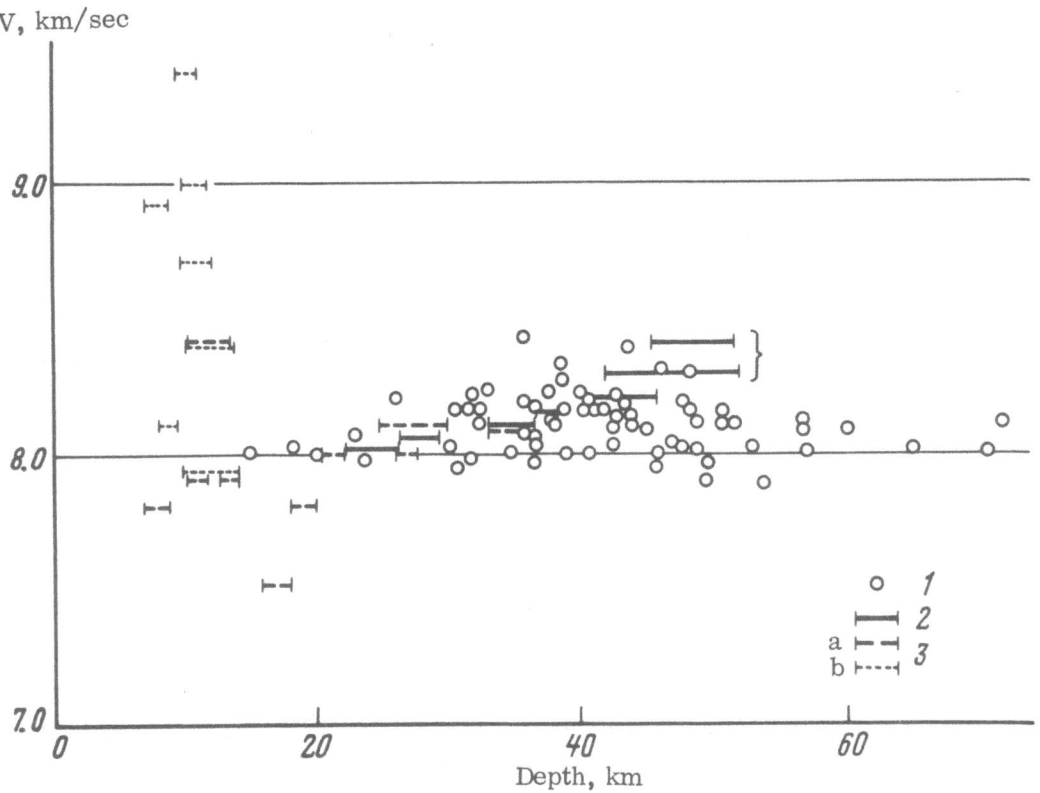

Fig. 6.3. Depths and velocities of the M-discontinuity: 1) continental crust [220]; 2) the same, from continuous profiling data within the USSR; 3) transitional zone between the Asiatic continent and the Pacific Ocean.

Examples of a complex, nearly continuous transition from crust to mantle in areas where a layer with a velocity of 7.4-7.6 km/sec is recognized, as well as a variety of velocities for the M-discontinuity are evidence that the blocky structure is present not only in the crust, but also in the upper mantle.

Examples of Blocky Structure of the Crust and Mantle. Blocks of various sizes in the crust and upper mantle are most obvious in the transitional zone between continents and oceans.

Crustal sections for the Sea of Okhotsk and the near-Kuriles zone of the Pacific Ocean are shown in Fig. 6.4. All of the deep seismic sounding data for the transitional zone have been thoroughly described in a book [163]; here, I will use them only as examples of a blocky crust with structures of various order.

Crust—mantle blocks with various types of crust are obvious on Section III in Fig. 6.4: continental (Sakhalin), near-oceanic (southern gulf of the Sea of Okhotsk), near-continental and continental (Southern Kuriles and the eastern continental shefl and slope), and oceanic crust. the Pacific Ocean.

Blocks are clearly evident within a continental crust on sections I and II in Fig. 6.4. These are characterized as regions of downwarp or uplift of the M-discontinuity, which are in accord with similar features for the surface of the crystalline part of the crust.

As has been indicated by detailed studies in the southern Kuriles region, in complex zones, blocks with smaller dimensions of about 30-60 km are also recognized, with the inhomogeneity characterizing either the entire crustal section or only part of it — upper or lower.

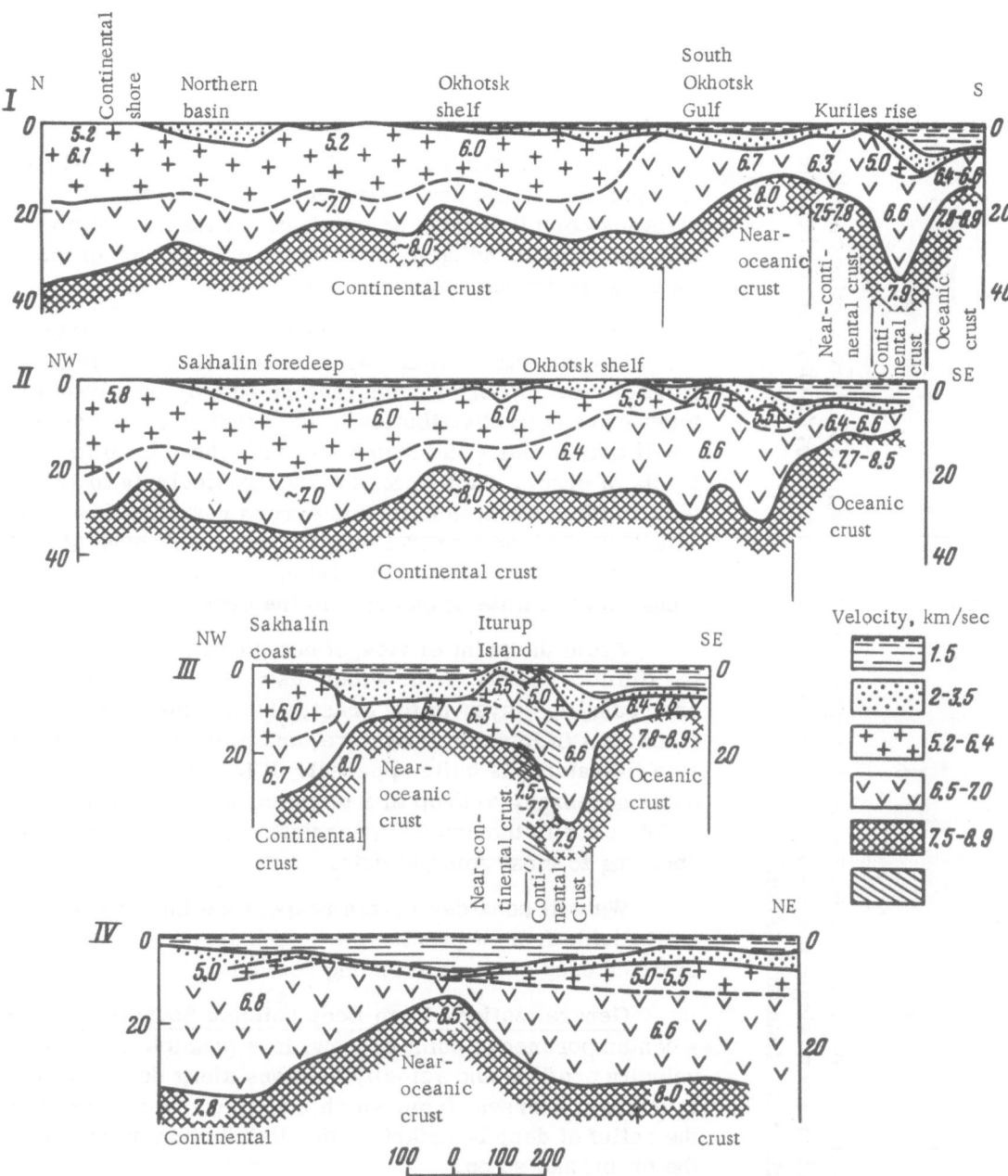

Fig. 6.4. Crustal sections [163].

The profiles from Zvenigorodka to Novgorod to Severskii, and from Temir-Tau to Petro-pavlovsk also provide examples of a blocky continental crust, as indicated on Figs. 4.32 and 6.5, respectively. Figure 6.5 is of particular interest, because blocks are recognized in a non-layered crust.

It is probable that there are similar nonuniformities and blocky-layered crustal sections for the oceans and oceanic plates. This is indicated by the relief map of the Pacific Ocean bed compiled by G. B. Udintsev and others [171, 172]. There are also seismic data indicating a blocky structure for the surface of the mantle east of the southern Kuriles (Tulina, [170]).

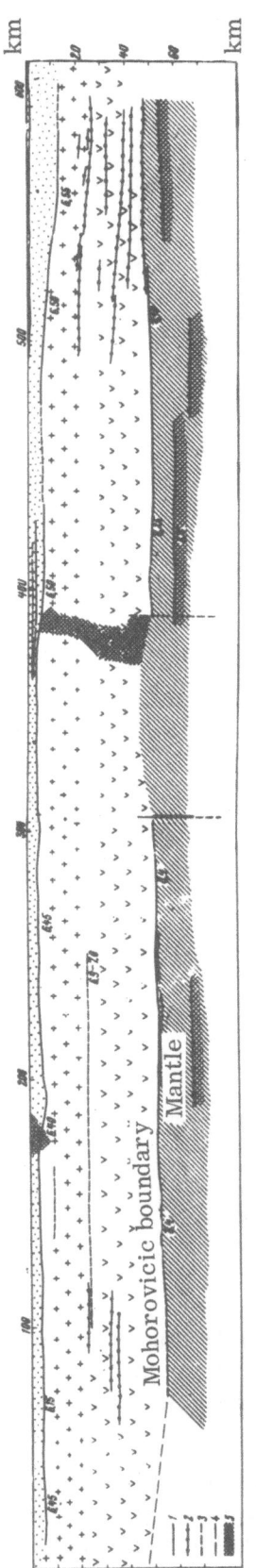

Fig. 6.5. Crustal section along the profile from Temir-Tau to Petropavlovsk (Antonenko [8]): 1) refracting boundary; 2) reflecting boundary; 3) less reliable boundary (from single travel-time curves, poor correlation, etc.); 4) interpolated boundary; 5) fracture zone. The numbers on the sections are boundary velocities.

Bullard [240] has noted some anisotropy in the properties of the M-discontinuity close to the Hawaiian Islands, which are possibly an example of a structural inhomogeneity similar to those discussed earlier in Chapter V.

Consideration of these examples shows how complex the structure of the crust and upper mantle may be in different tectonic zones, and how over-generalized the initial characteristics compiled from seismological data and from the early deep seismic sounding data, in which only groups of mantle waves were traced, must be.

Data on the intermediate boundaries on the early deep seismic sounding sections, when viewed from our present point of view, apparently characterized only the gross features of velocity distribution in the crust, while structures of small amplitudes (less than 5 km) must have been lost in the limits of error. Such errors arise, particularly in relation to the wide distribution of time errors resulting from the identification of strong, predominantly reflected waves at high angles from the M-discontinuity, as head waves refracted along intermediate boundaries in the crust.

From the point of view of complex layering in the crust, the limited validity of the widely used concepts of a "granitic" and "basaltic" layer in the crust, which unfortunately have taken root in Russian literature, and even in some non-Russian literature, is quite apparent. It is obvious that it has become necessary to drop this terminology [177], which has been widely used in the general description of the concepts corresponding to most seismic data.

We will consider in this respect the information which is available from a precise physical interpretation of generalized deep seismic sounding sections.

Generalizations from Deep Seismic Sounding Data. From a contemporaneous point of view, it is possible to construct velocity profiles and velocity sections along deep seismic sounding traverses, from which maps may be prepared showing the relief of deep boundaries, the distribution of velocities in the crust, and so on.

Seismic sections contain two forms of data: seismic boundaries — refracting and reflecting — which correspond to regions with a more or less abrupt change in properties of the medium and data on the layer and average velocities, which provide an idea about the general relationship of velocity with depth and with horizontal distance. The parameter defining a seismic refracting interface is the boundary velocity, which characterizes a reasonably thin layer with no recognizable increase in velocity with depth in the layer (a head wave), or a thicker layer in which the velocity increases significantly with depth between boundaries, beginning at the boundary (a weakly refracted wave).

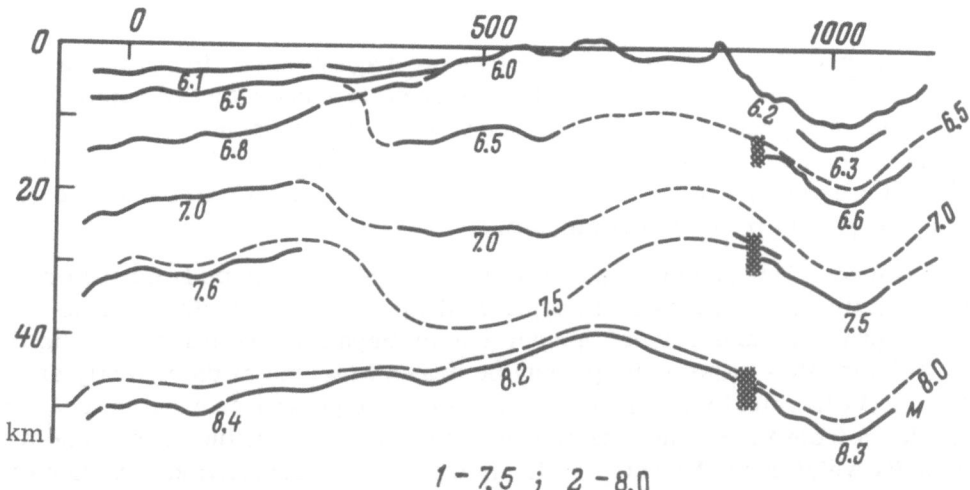

Fig. 6.6. Contours of a velocity field (from [42]). Solid lines are seismic boundaries and the corresponding numbers are boundary velocities. The dashed lines are velocity contours determined by interpolation between boundaries.

As we have seen, for deep seismic sounding, seismic boundaries are usually discontinuous, with changes in properties (refracting and reflecting) along a traverse, and their identification between unrelated regions is difficult. On the average, the most reliable feature in the block of a single type is the velocity profile, reflecting the general relationship between velocity and depth. In particular, the general character of the seismic wave field is related to the velocity profile. In this respect, it would be desirable to supplement the usual sections for reference seismic boundaries with contours of layer velocities (that is, higher than or lower than true values).

This manner of presenting data has not yet been widely used because of the difficulty involved in determining velocity. Several methods for constructing velocity fields are available, based on the use of various types of data (Litvinenko [116], Pushkarev et al.). I will not evaluate them here. I will only give an example of such a field as an illustration of its usefulness.

A velocity field is shown in Fig. 6.6 as contours constructed from a deep seismic sounding section, on which the boundary velocities are indicated [42]. It was constructed on the assumption that the boundary velocity corresponds to the upper half of the layer beneath the boundary, while in the lower half of this layer, the velocity increases linearly from V_k to V_{k+1}, where k is the index number for the upper boundary and k + 1 is the index number for the lower boundary. The contour interval is 0.5 km/sec.

The field clearly outlines the principal features of the section that we usually use, and in particular, the presence of three crustal blocks with differing velocity levels.

Against the general background of these blocks expressed as an undulation of the contours, the seismic boundaries are clearly seen in detail. It is quite obvious that boundaries situated in different blocks at essentially the same depths correspond to different velocity levels (boundaries with 7.6, 7.0, and 7.5 km/sec), i.e., in the various blocks, the surface with sharp changes in the physical properties of the medium are situated in different rocks with respect to composition and structure.

Thus, the behavior of velocity contours and seismic boundaries makes it possible to analyze simultaneously two factors: first, the general distribution of velocity with depth, which characterizes the smooth variation in space of the average composition of the material in the crust, and second, regions of relatively sudden changes in these average relationships, which reflect the location and properties of the primary seismic boundaries (velocity boundaries). It is obvious that these data must also serve as the basis for generalized deep seismic sounding sections. Contoured velocity cross sections also are good indicators of low velocity zones, which may be recognized from reflection data.

We will consider how it may be possible to prepare areal maps from the properties of the velocity cross sections, using data from adjacent deep seismic sounding profiles. Apparently, there are two approaches to constructing such maps: maps may be drawn showing the depth to a specific velocity V_i = const or maps may be drawn in terms of the velocity at a specific depth H_i = const (V_i is velocity and H_i is depth). Either type will provide an idea about the proximity or depth from the surface (or the surface of the crystalline part of the crust) for a specified velocity which may characterize the petrographic composition of various rocks.

Cross sections and contour maps of velocity may serve also to compile maps or sections with pressure and temperature contours (see Fig. 6.11).

The combination of these data on formation velocities and their comparison with geological structure and history in the region under study may give some idea of the dynamics of such processes as metamorphism, which apparently is the reason for various seismic boundaries in the crust.

Greater detail, which is important for comparison with local structures on surfaces, and generally, greater accuracy may be obtained by areal mapping of the seismic boundaries – their relief or their properties (as, for example, boundary velocity). At present, this is quite commonly done. In addition to the primary boundaries (the K_0 and M), which may be traced reasonably easily even with data of only moderate reliability, intermediate boundaries in the crystalline crust may also be traced if sufficient data are available and if the systematic errors which were discussed in Chapter IV are considered. In so doing, however, serious difficulties are encountered in identifying such boundaries between unrelated profile segments, as may be seen from Fig. 6.6.

An example of such a difficulty is that of correlating the boundary with a velocity of 6.0 km/sec in the center of the diagram with the boundaries having velocities of 7.0 and 7.6 km/sec to the left and those with velocities of 6.6 and 7.5 km/sec to the right. In the general solution of this problem, and in this case in particular, it must be remembered that to trace a single seismic boundary reliably along a traverse, it is necessary to correlate related sets of data, or have other criteria for identifying the various waves recorded along the separate profile segments. Usually this is not possible with deep seismic sounding data. However, each seismic refracting boundary on a deep seismic sounding section corresponds to a specific parameter, i.e., the boundary velocity, which characterizes the velocity for an average crustal section in a given region.

Inasmuch as the variations in velocity for a single boundary at about the same depth are large, it is apparent that the most objective approach may be the construction of cross sections on which individual velocity values are noted as tie points for velocity contours. Given reasonably tight limits on these velocities, we may follow the change in depth of the contours in various tectonic zones; that is, we may study the relief of velocity zones. In some cases, seismic boundaries may be present in these depth intervals, and in other cases, not – and this will be indicative of how the properties of these velocity zones vary in the crust.

Thus, in analyzing deep seismic soundings or comparing them with other seismic data, it seems to us to be quite irrelevant to resort to schematic crustal sections consisting of two layers only, with broad limits to the velocities which carry the information about layer properties. Obviously, it would be more satisfactory to develop methods of analysis based on a more complete utilization of the properties of the velocity section and the discrete boundaries within it — refracting and reflecting boundaries.

§ 2. Average Velocity Characteristics for the Crust and the M-Discontinuity

Keeping in mind the development of approaches to typifying the crust, it will be of interest to review new data on the statistical characteristics of velocity profiles in the crust for various morphological zones in the continents and oceans.

The majority (more than 75%) of the 1200 crustal velocity profiles which have been determined pertain to the oceans [223]. Over these wide areas, velocity profiles have been calculated from single travel-time curves, obtained over short profiles (lengths of about 100 km). Thus, in actuality, they characterize average values for velocities and depths at specific points.

During the International Geophysical Year, detailed surveys along long traverses were carried out in a number of oceanic regions, which permit the construction of summary crustal sections along profiles crossing various types of structure [207, 232, etc.]. The total length of deep seismic sounding profiles throughout the world amounts to 200,000 km.

Studies of crustal structure along long traverses in the USSR are of particular interest. The distribution of deep seismic sounding profiles over the USSR and the nearby oceans is shown on Fig. 6.7.

We will consider the average statistical data on the velocity makeup of the crust in various regions. Unfortunately, not all the data from the USSR are yet included.

Velocity in the Earth's Crust. Statistical reduction of velocity profiles [223] obtained from refraction data gives an idea about the distribution of three types of seismic velocities in the crust and upper mantle: 1) layer (or boundary) velocities V_p for specific layers in the crust; 2) average velocity $\overline{V}^M$ for the entire solid crust down to the M-discontinuity (excluding the water layer in seas and oceans); 3) the velocity V^M in the mantle directly beneath the M-discontinuity (Fig. 6.7).

Figure 6.8 which shows the average (with respect to depth) velocity, $\overline{V}^M = \Sigma (V_i h_i)/\Sigma h_i$, gives an idea of the velocity values that are dominant in the sections.

Inasmuch as the number of determinations of velocity in the oceans is nearly three times greater than the number on dry land, the curves for the earth as a whole are determined practically by the character of the oceanic curve.

There are a number of significant differences between the oceanic and continental curves. In considering the distribution curves for the formation velocities V_p it may be seen that velocities of 1.5-2.5 km/sec corresponding to the sedimentary section in the oceans and of 5.5-6.5 km/sec corresponding to the upper part of the crystalline crust on the continents are met most frequently in the crust. The differences between the curves for the oceans and continents appear in the relative sizes of the peaks for the continents; the diagnostic peak is at high velocities, while for the oceans it is at low velocity. This difference reflects, in our opinion, primarily the inadequate amount of data for sedimentary rocks on the continents, while many determinations have been made for the upper part of the section for the oceans.

The average velocity in the crust $\overline{V}^M$ differs significantly for the continents and oceans: it is nearly 1 km/sec smaller for the oceans. This is explained by the larger role played in

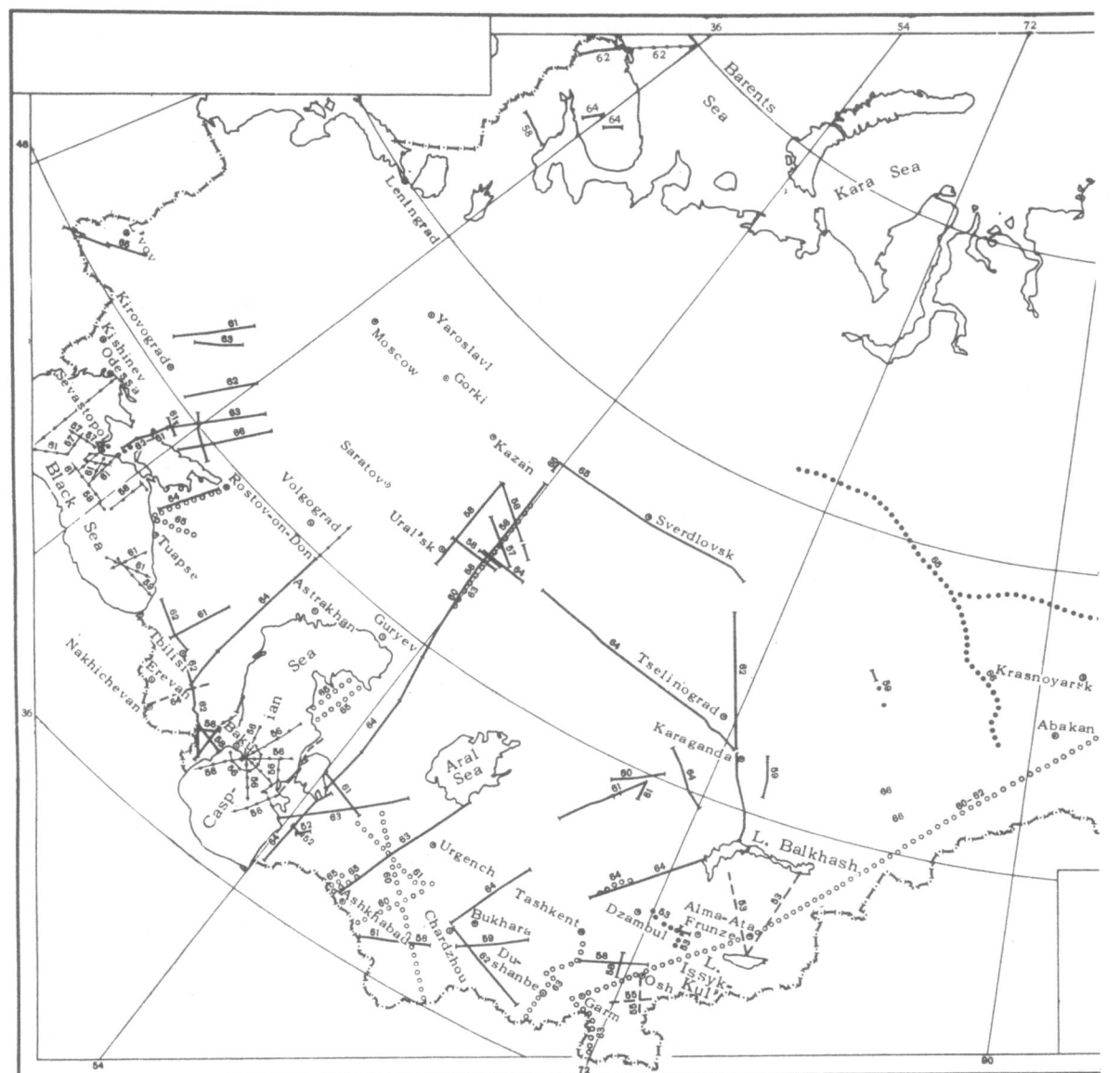

Fig. 6.7. Deep seismic sounding profiles in the USSR (data from VNII Geofizika, Vol'vovskii
4) point sounding; 5) seismological

oceanic sections by the low-velocity sedimentary layer, comprising about one-fifth of the total
thickness of the crust. It is characteristic that for the continents, the average velocity in the
crust is apparently at least 5.4 km/sec. In some very rare cases [3], when a thick sedimentary section is present, the average velocity may be somewhat lower, down to 4.5 km/sec.

The distribution curves for the velocity at the M-discontinuity V^M are interesting. In
[219], all boundaries with a velocity greater than 7.6 km/sec are defined as being the Moho
boundary. Distribution curves for the velocity V^M are similar to one another for both oceans
and continents, and show a narrow maximum in the range from 7.9 to 8.3 km/sec. For the continents, the majority of determinations give 8.1 to 8.2 km/sec; for the oceans, 7.9 to 8.6 km/sec.
These curves are completely compatible with those presented earlier in Fig. 6.3.

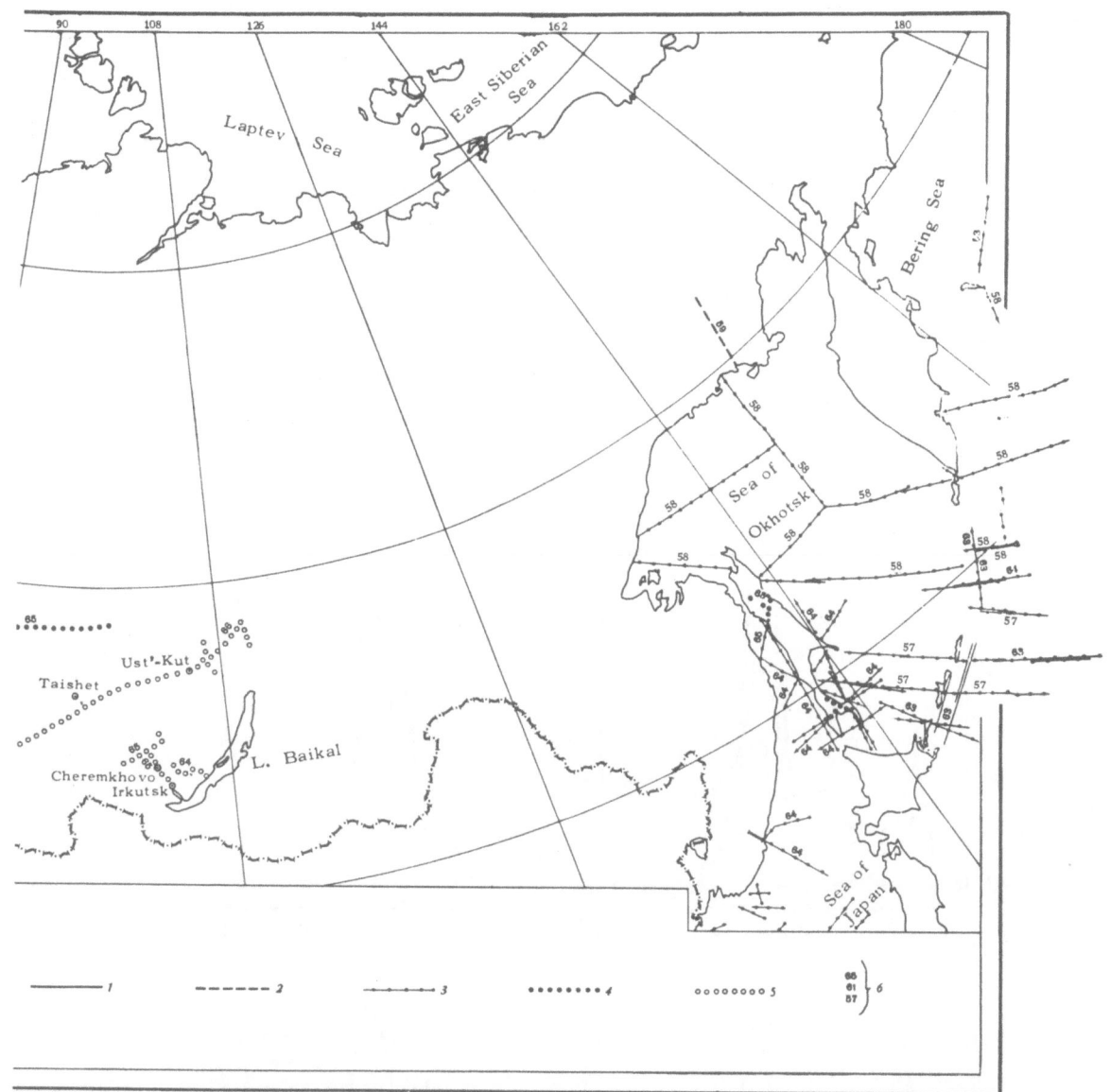

et al., 1966): 1) continuous profiling; 2) segmentally continuous profiling; 3) point profiling; profiling; 6) year work was done.

Statistical data on velocities for several structural zones are given in Table 23. There is a small difference in average velocities in the crust for structures which are similar in type.

A comparison of similar structures on the continents and oceans as for example, oceanic plates and continental shield areas, underlines the differences which were discussed earlier in evaluating the curves for the earth as a whole.

Reference [220] gives average velocity sections for various types of structures, which are shown in simplified form in Fig. 6.9.

Departures from the average data generally are diagnostic of specific structures. However, in each specific case, these may frequently be explained by the approximations which

TABLE 23. Velocities and Crustal Thickness

	Average	Continental shelves	Shields and stable platforms	Inactive Plaeozoic mountain areas	Active mountain areas	Plates and mountains	Islands and ridges	Average	Continental inland seas	Mid-ocean ridges	Oceanic areas
Elevation above sea level, km	0.53	0.06	0.85	1.3	1.16	1.46	0.15	−2.02	−1.32	−3.72	−4.4
Velocity in the crust, km/sec	6.24	6.04	6.34	6.10	6.40	6.41	6.73	5.56	5.43	5.45	5.67
Depth to Moho, km	40.13	31.33	44.06	44.43	41.06	48.34	27.80	9.64	10.72	6.29	9.05
Velocity at M-discontinuity, km/sec	8.09	7.99	8.16	8.10	8.05	8.08	8.00	8.11	8.09	8.03	8.13
Number of determinations	175	20	30	24	17	5	3	320		23	148

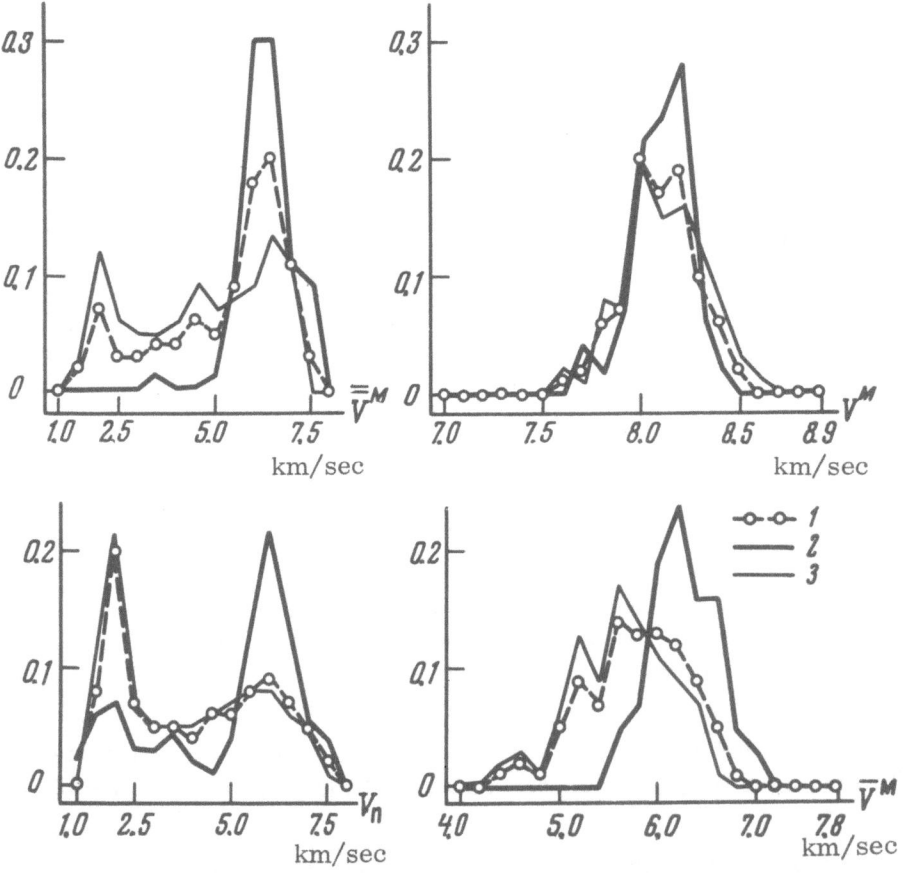

Fig. 6.8. Distribution of velocities in the crust: 1) for the earth in general; 2) for the continents; 3) for the oceans; V_p layer velocities; $\overline{V}^M$ average velocity in the crust down to the M-discontinuity; $\overline{\overline{V}}^M$ average interval velocity in the crust; V^M velocity at the M-discontinuity. The vertical axis indicates that number of determinations of specific velocity values, expressed as a fraction of the total number of determinations.

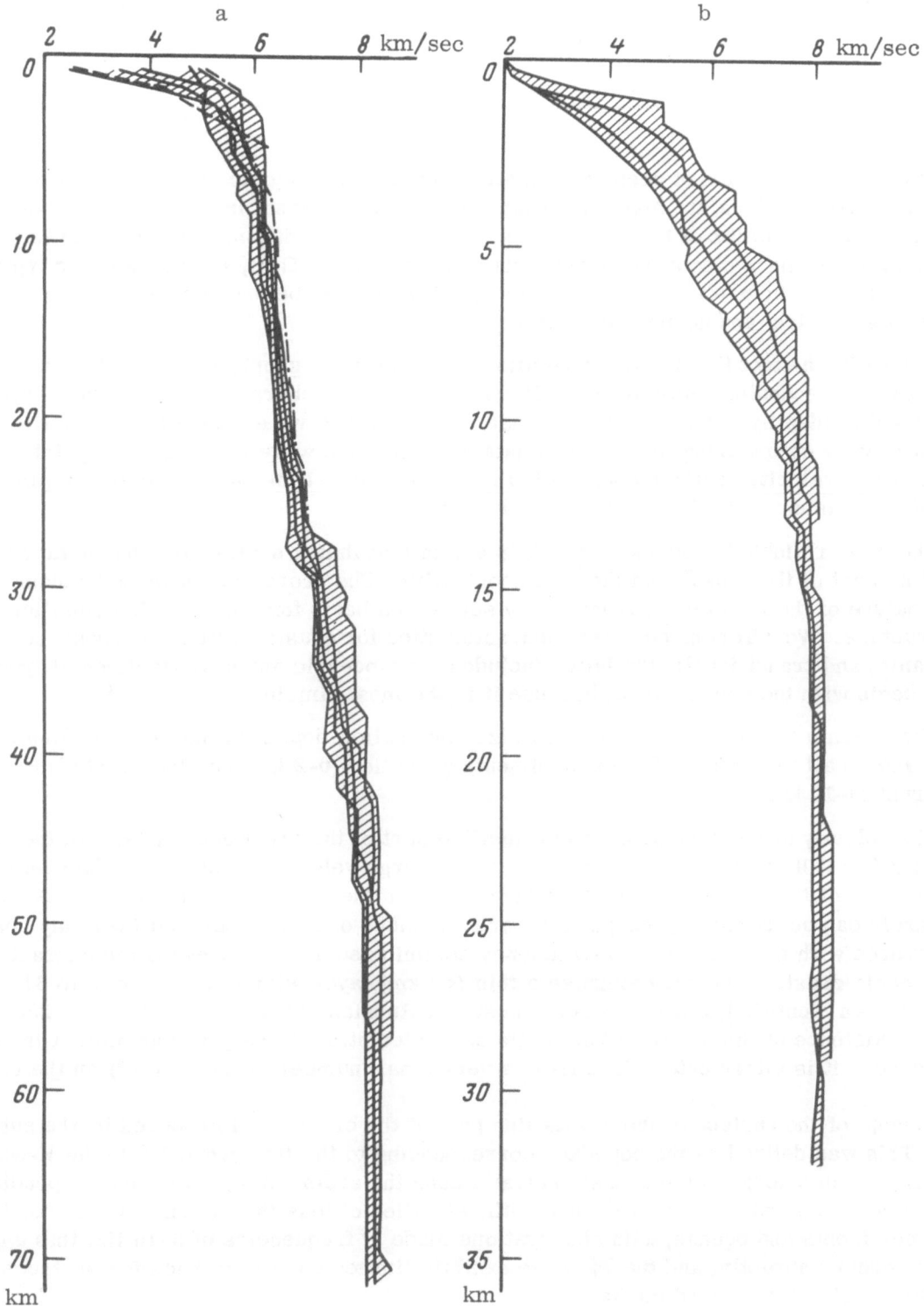

Fig. 6.9. Relationship of velocity to depth: a) for continental structures; b) for oceanic structures.

were made in interpreting the travel-time curves. The extent of approximation used depends primarily on the degree of completeness of the data, a matter which was discussed in §1 of this chapter.

§3. Types and Structures of the Crust of Various Orders

Extensive application of deep investigations in various tectonic zones has indicated that variations in type such as are apparent from the examples present in §§1 and 2 correspond to specific characteristics of the deep crustal structure. In addition, similar features in the upper mantle have been discovered over rather broad zones. Thus, a real basis for typifying the deep crustal structure is available. This typification may be based on a variety of indicators of geological and geophysical nature.

In view of the fact that the interpretation of virtually all geophysical data is now based on deep seismic sounding sections, we will consider how the characteristics of these sections may be used to identify various types of crust. In Chapter V, we evaluated this question from the point of view of analyzing the primary data — the seismic wave field. At this point, we will again consider velocity sections constructed from the travel-time curves for the dominant elements of the seismic wave field — the basic wave groups.

The most reliable boundaries on a deep seismic sounding section are the surface of the crystalline part of the crust[†] and the M-discontinuity. The properties of these boundaries and the nature of their relief apparently may serve as a basis for the seismic classification of deep structure. We will consider their characteristics for crusts of various types: continental, oceanic, and transitional; the latter includes near-oceanic and near-continental types. We will begin with the oceanic type, because it is the most simple.

The oceanic type of crust is characterized by shallow depths to the M-discontinuity (15-20 km), and small thicknesses for the sedimentary section (0-2 km) and the crystalline part of the crust (4-10 km).

The velocity at the surface of the crystalline part of the crust usually falls in the range 6.4-7.0 km/sec. It increases slowly with depth. A large velocity gradient may be present in the lower part of the crust. Intermediate boundaries in the crystalline part of the oceanic crust rarely can be identified, though there are a number of indications that they may possibly be discovered with the use of higher-frequency seismic records. Reviewing the characteristics of the oceanic crust, we do not recognize a thin (<2 km) layer with a velocity of 5 to 6 km/sec which has been identified in the oceans by most non-Russian authors [143, 207, 209, 228, 229]. The very existence of this layer, let alone its characteristics, is very problematical in many cases because it is either determined from a very small number of points (1-2) on the travel-

---

[†] The concept of the surface of the crystalline part of the crust was introduced by the author in 1962. This was defined as the boundary corresponding to the first wave P_0^K in the K-wave complex, which emerges as the first arrival among the sedimentary waves corresponding to boundaries in the sedimentary sequence with velocities of less than 5 km/sec. In most areas on the continents and oceans, with observations made at frequencies of 5-15 Hz, this emergence occurs abruptly, and the P_0^K wave exhibits limited characteristics for surfaces in the so-called vertically layered media.

However, in complicated areas, there is a transition from sedimentary waves to K-waves, and the P_0^K event may be assigned to a well-defined boundary in the lower part of the sedimentary sequence. The geological evaluation of this term will be discussed in §4.

time curve for first arrivals or determined by tracing very roughly later arrivals on the records. On the basis of data from the northeastern part of the Pacific Ocean, this velocity cannot be reliably found, which may be related to the low density of observations at short distances from the source. Velocities of less than 5 km/sec may be referred to the sedimentary section. In this sequence, finally, thin layers with higher velocities may be present, such as were found in the Phase I Moho drilling near Guadalupe Island, in the case of an intrustive basalt layer (Bullard [192-194]).

The velocity of the M-discontinuity in areas with an oceanic crust may range over the limits – from 7.8 to 8.6 km/sec and higher [170, 220]. The phenomenon of structural anisotropy which correlates with magnetic anomalies, has been observed [170, 240].

The continental type of crust has been fully characterized in § 1. The basic indicators are a relatively great depth to the M-discontinuity (20 to 75 km) and, consequently, a greater thickness to the crystalline part of the crust. The sedimentary section may vary in thickness (0-10 km, and more). The velocity at the surface of the crystalline part of the crust is close to 6 km/sec. The velocity increases with depth. Several intermediate boundaries have been discovered in the crust. At moderate depths (10 to 30 km), these boundaries are characterized by velocities of 6.5 to 7.0 km/sec, while at greater depths, by velocities of 7.0 to 7.6 km/sec (see Fig. 6.7). Significant deviations in the velocity characteristics of the crust have been discovered in some deep basins (see Fig. 6.2).

A comparison of the seismic sections for continental and oceanic types of crust indicates that the velocity in the oceanic crust essentially corresponds to the velocity in the middle part of the continental crust. In this respect, an analysis of the differences between these two primary crustal types indicates that there is some basis to the concept which arises from a two-layer model of the crust that there is no "granitic" layer under the oceans with a velocity of 6 km/sec; this is so, as may be seen readily from Fig. 6.10, and the lower layer of the continental crust has a velocity greater than 7 km/sec.

On Fig. 6.10, the velocity profiles represent sections for the mantle under continents (from deep seismic sounding sections [152]), island arcs (from seismological studies by S. A. Fedotov et al. [175, 176]), and the oceans. These illustrate layering in the upper mantle similar to the layering in the crust, and the suppression of this layering or possibly its complete disappearance in the mantle under the oceans. These new features of the detailed velocity sections for the uppermost part of the mantle, directly under the crust, are essentially supplementary to the earlier seismological concepts, so that there is no effect on hypotheses concerning the relationship between the crust and mantle or on the reasons for the differences between the continents and the oceans.

We will consider the two transitional crustal types further: the near-oceanic type and the near-continental type.[†] The characteristics of these velocity sections, as has already been indicated in the analysis of seismic wave fields in Chapter V, are less specific than in the case of the primary types.

The near-oceanic type of crust is distinguished from the oceanic type essentially only by the greater thickness of the sedimentary section. The characteristically thin crystalline crust in deep-water basins of marginal seas is basic [86, 99, 223, etc.]. The somewhat thicker crust of deep-water basins in inland seas is also diagnostic.

---

† Both terms (near-oceanic and near-continental) were introduced by Kropotkin (see, for example, [58]). The use of these terms has been criticized by some geologists and geophysicists. Even so, we use the terms because in our opinion they stress the relationships between the principal types of crust and, to a secondary extent, the distribution of these types.

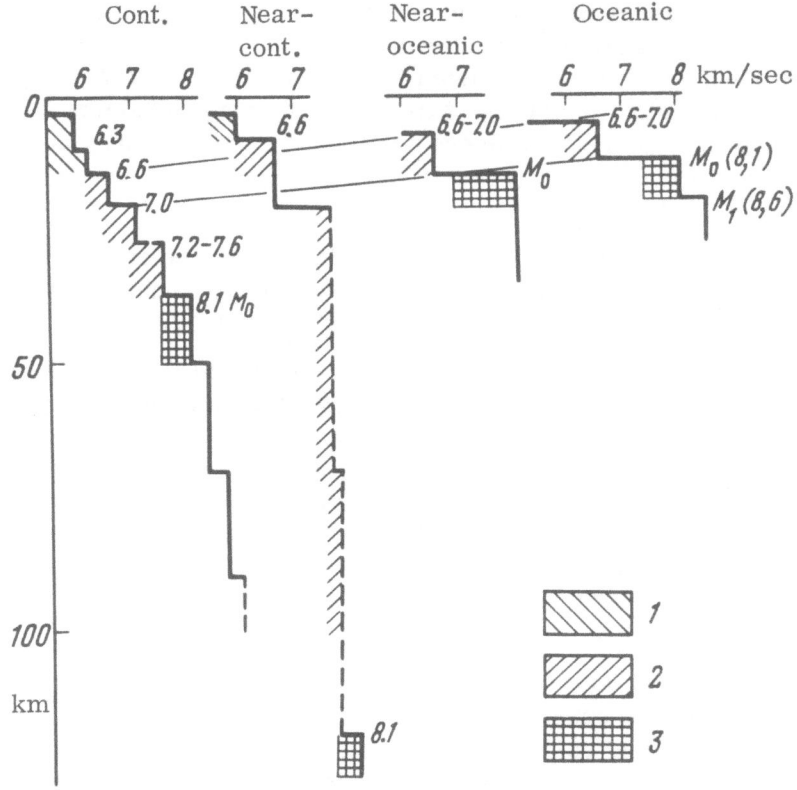

Fig. 6.10. Velocity profiles through the crust and upper mantle: 1) velocity of less than 6.6 km/sec; 2) more than 6.6 km/sec but less than 8.1 km/sec; 3) velocity of 8.1 km/sec or higher. Continental crust from Ryaboi [152]; near-continental from [163] and Fedotov [175, 176]; near-oceanic from [163]; oceanic from Zverev and Kosminskaya [109].

However, one may think that though the upper part of the crystalline part of the crust under inland seas has the same velocity as in the oceanic crust, the lower part of the crust in such basins as the South Caspian apparently has properties characteristic of a continental crust. This hypothesis is based on a comparison of the seismic wave fields recorded in the southern basin of the Caspian Sea, where no layer having a velocity of 6 km/sec has been discovered, and in the Prikurin Depression [64], where this layer is very thin or completely absent, but a layer with a velocity of 6.8 to 7.2 km/sec is present beneath (see Fig. 6.7).

In this respect, it is not completely clear which part of the Black Sea basin is an area with a near-oceanic crustal character [124]. Probably in the deepest parts, where the thickness of the crystalline part of the crust is less than 10 km, it actually has a velocity section similar to that in the near-oceanic crust of the bordering seas. However, in other areas nearer to the coastline, where the thickness of the crystalline part of the crust increases to 15-20 km, its lower part probably assumes velocity features characteristic of the middle and lower parts of a continental crust.

Thus, a near-oceanic crust has a small depth to the M-discontinuity (up to 15-20 km), a thin crystalline crust (3-15 km), oceanic velocities in the upper part, and a sedimentary section which is comparable in thickness to the crystalline part of the crust.

The near-continental type of crust represents sections with relatively small depths to the M-discontinuity (< 25 km) and a moderately thick crystalline part of the crust (more than 15 km), in which the upper part has the characteristics of a continental crust; that is, the velocity is about 6 km/sec at the basement surface. Examples of crustal structures of this type are apparent on deep seismic sounding sections from the Kuriles. The thin crystalline crust in the Hungarian uplift, where the velocity is predominantly around 6 km/sec, while a boundary with a velocity of 6.6 km/sec is present only in the lower part [233, 234]. It should be noted however, that data on the structure of the lower part of the crust in Hungary are not adequately detailed, and the interpretation may be subject to significant changes.

Areas in deep-water basins of marginal seas, where a marked thickening of the crystalline part of the crust takes place should apparently be assigned to the near-continental class, as noted earlier.

If we examine crustal velocity profiles sequentially from the ocean to the continent, we may see the following. In the transition from oceanic to near-oceanic crust, the thickness of the crust increases because of the sedimentary section, while in the further transition to near-continental and continental crust, it increases because of the appearance of layers with velocities of 6 and 7-7.6 km/sec. There is practically no change in the middle part of the crust, which has a velocity of 6.6-6.8 km/sec. Thus, we may speak of two causes for crustal thickening — by the addition of a sedimentary sequence to its surface, which then is subjected to metamorphism and tectonogenic processes to be transformed into crystalline rock, and by the addition of crust from beneath by differentiation of the mantle and the separation of its lighter components. An attempt to explain these processes from the geological and petrographic point of view has been formulated most clearly in the later papers of V. V. Belousov [17, 19]. We will consider these further in the following paragraphs.

Crustal Structures of Various Orders. An important capability of deep seismic sounding is that of studying the relief of a seismic boundary along a traverse. This permits classification of crustal structures not only with respect to thickness and velocity characteristics, but also according to the form of the structure in a plane, as well as in relation to the relief of the principal deep boundaries.

The combination of the velocity characteristics of the upper part of the crystalline crust and the nature of the relief on the basement surface (the K_0 boundary) and the bottom of the crust (the M-discontinuity) provides a basis for the subdivision of blocks into crustal structures of lesser dimensions of various orders [101, 218].

The characteristics of structures of various orders for the transitional zone between the Asiatic continent and the Pacific Ocean are listed in Table 24.[†]

The primary features in the combinations of various forms of relief are: the concordance of the two bounding surfaces for the continental crust in the northern and central part of the Sea of Okhotsk and the marked discordance of these boundaries for the south Kuriles Gulf, the Okhotsk Gulf, and the Komandor Gulf of the Bering Sea (see Fig. 6.4).

In the junction zones of various types of first-order structures, there are marked changes in slope of both boundaries.

This system for classifying structures according to seismic data must be supplemented with other geophysical data: characteristics of the gravitational and magnetic fields, heat flow, and seismological evidence of the present activity of the crust and blocks. Such a generalized geophysical classification for the earth's crust, compiled considering all the reliable geo-

---

[†] This classification was developed by the author in collaboration with Yu. V. Tulina, S. M. Soerev, and R. S. Veitsman (see [163]).

TABLE 24. Classification of Crustal Structures

Order I	Order II (thickness of crust	Order III*	
		region	structure
Continental crust	(25-80 km) Continents and continental shelves	Profile from Magadan to Kolyma, northern part of the Sea of Okhotsk, western slope of the Kurile-Kamchatka rise	West Okhotsk basin, south Kuriles basin, north Kuriles basin
	(20-25 km) Shallow-water areas of inland seas	Northern and central part of the Sea of Okhotsk	Okhotsk rise, north Kuriles plateau
Near-continental crust	(15-20 km) Island arcs, oceanic islands	Southern part of Kurile Island Arc	South Kuriles plateau
Near-oceanic crust	(10-15 km) Marginal seas (15-25 km) Basins of inland seas[†]	Southern part of the Sea of Okhotsk, Bering Sea	Southern Okhotsk basin, central Kuriles basin
Oceanic crust	(10-15 km) Oceanic deeps (10-15 km) Oceanic plateaus	Northwest part of the Pacific Ocean	Pacific basin

*Structure of III order are given for the transitional zone between Asia and the Pacific Ocean [163].

†From a contemporaneous point of view, the thick crust in the deep-water part of the Caspian Sea should more properly be assigned to the continental type, apparently.

physical data, may then be related to the geological factors which lead to the recognition of crustal structures of various types, and each deep geophysical structure will acquire specific geological meaning.

It seems to us that such a system for classifying geophysical data will be very important. Without such a classification, interpretation is to a large degree uncertain and inexact, inasmuch as different types of data are used for identifying various types of structural combinations.

At present, there are several schemes for classifying crustal structures on the basis of combinations of geophysical data. Summaries have been published recently by G. Z. Gurarii and I. A. Solov'eva [58] and by McConnell and Taggart-Cowan [223]. A classification of deep structures in the USSR has been given by A. A. Borisov [27].

It is clear that all of the new data bears a somewhat "disordered" relation to the structure of earlier classification systems. In a specific case, the author has attempted to relate these to the accepted concept of a "granitic" and a "basaltic" layer in order to make the concept compatible with the new seismic data on the multiple layered nature of the crust and differences in the characteristics of the transition from crust to mantle in areas with a thick continental crust and a thin oceanic crust.

§4. The Nature and Cause of Layering in the Crust and Upper Mantle

Contemporaneous geophysics and geology provides much direct and indirect information on the composition and nature of the material in the solid earth, which has not yet been adequately evaluated.

We will examine the seismic data on the elastic properties of the crust and mantle from the point of view of correlation of such information with similar parameters for rocks at high pressures and temperatures. In so doing, we will also keep the principal feature of the crust and mantle in mind — the velocity layering.

There are many data in both the Russian and non-Russian literature on laboratory determinations of the elastic properties and densities of rocks with various petrological compositions both under normal pressures and at high pressures and temperatures [13, 40, etc.]. These data are so abundant that they may be analyzed statistically.

Measurements of the velocity and density of samples of acidic, intermediate, basic, and ultrabasic rocks with various compositions have indicated that there are significant differences for averaged data but individual values for samples in each group may overlap with values for a neighboring group [189, 190, 211]. Having a combination of data on the physical properties of various igneous and metamorphic rocks, including metamorphosed sediments (limestone, sandstone, marble, dolomite, and others) allows us to evolve ideas about the laws for relationships between compositions, densities, and velocities. The main features are expressed roughly as an increase in velocity and density as rocks become more basic [13, 182]. It is also known from tectonic and geochemical considerations that rocks become more basic with increasing depth (Belousov, Sheinmann [19 and 186, respectively]).

There may be significant exceptions to these laws for specific groups of rocks or specific samples of each rock type. There may also be exceptions to the depth relationship as a consequence of conditions involved in the development of various tectonic zones [142].

At the high pressures and temperatures characteristic of the crust and upper mantle, the general relationship between velocity and the chemical nature of a rock is preserved. The velocity for each rock type increases with depth of burial. The lower the velocity and density of a rock are, the more rapidly will the curve relating velocity to pressure rise (Fig. 6.11). As has been shown by the recent work of M. P. Volarovich and others [40, 41], this behavior persists to pressures of 10-20 kilobars (approximate for depths of 30-35 km to 60 km in the earth). At higher pressures, the velocity begins to increase again.

The increase of temperature with depth exerts the inverse effect, frequently compensating the effect of the increase in pressure (Fig. 6.12). Thus, when the effects of temperature are considered, at depths more than 20 km, the curve for the relationship between velocity and pressure for a given rock type becomes almost parallel to the pressure axis. At great depths (greater than 100 km), according to Clark and Ringwood [201] the effect of temperature may override the effect of pressure [75], with the result that the well-known low-velocity layer of Gutenberg is formed. Thus, we presently have adequate experimental data to evaluate the way in which velocity varies with depth under normal conditions of pressure and temperature for an assumed composition of the crust and upper mantle.

If the crust were to consist of rocks of single composition, the velocity would increase with depth as shown by the curves in Fig. 6.12. However, the distribution of velocities in the crust of the continents and oceans is actually more complicated. As shown by the results of deep seismic sounding, velocity does not increase uniformly with depth, but discontinuously, and the rate at which the velocity increases at depths of more than 5 km is higher than that for any of the samples used in laboratory studies. This is clearly evident from Fig. 6.13, where the ranges of velocities for rocks in the granite and gabbro-diorite classes are shown superimposed on the ranges of velocities at seismic boundaries in a continental crust, as well as the typical sections for continental and oceanic sections of the crust and upper mantle given in Fig. 6.10.

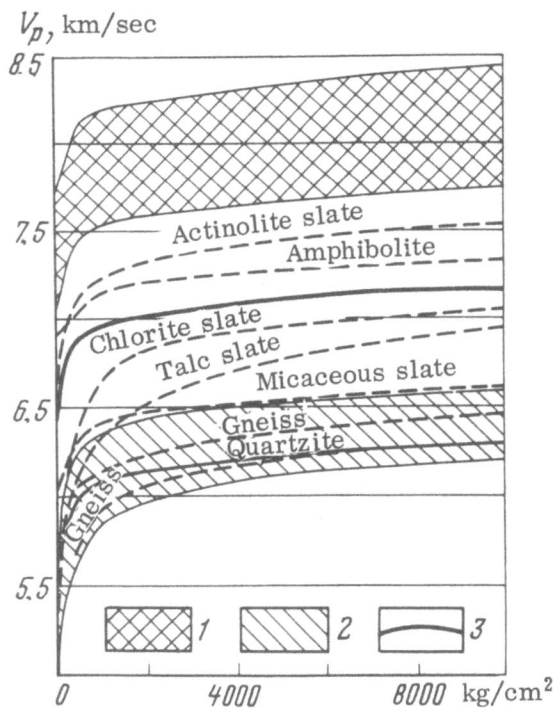

Fig. 6.11. Relationship between velocity and pressure for igneous and metamorphic rocks [190]: 1) dunite; 2) granite; 3) gabbro.

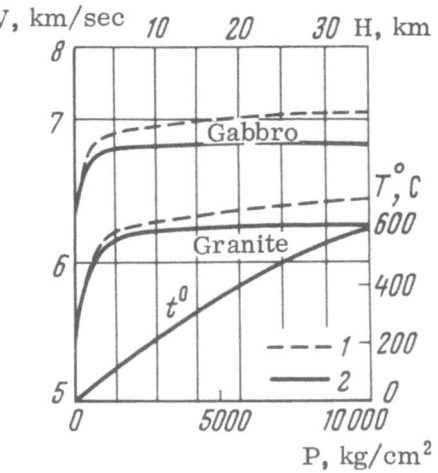

Fig. 6.12. The effect of temperature on the variation of velocity with pressure [190]: 1) at 18°C; 2) with consideration of the increase in temperature with depth.

Comparing these data, we should consider the following features. First, in a continental crust, the velocities at depths of 0-10 km correspond to those for acidic rocks; at depths of 10-30 km, they correspond to velocities for intermediate and basic rocks, and in the upper mantle, to the velocities for ultrabasic rocks; in an oceanic crust, the topmost layer has velocities corresponding to those for rocks of intermediate and basic composition.

The second feature is the fact that if we consider each seismic boundary to be a reasonably sharp change in the elastic properties of the medium over a short interval in depth, then in comparing data from laboratory measurements with velocity profiles from deep seismic soundings, we may assume that the seismic boundaries correspond to contacts between rocks with differing compositions [142]. There is no difficulty involved in selecting rock types, as indicated earlier, at these discontinuities which will correspond to any of the data.

It is clear, however, that such an approach to the analysis of seismic data satisfies neither the geology nor the geophysics, but it does provide a conditional basis for calling layers "granitic" or "basaltic." However, in order to judge what types of rocks comprise the crust in a given area, it would be necessary to use other methods for comparing laboratory and experimental data on velocities. These apparently must be based on the study of the distribution of various rock types in the tectonic zones where deep seismic sounding profiles have been completed.

A mass of determinations of the elastic constants for the rocks which dominate in a given region is necessary for comparison with deep seismic sounding data, particularly for the rock

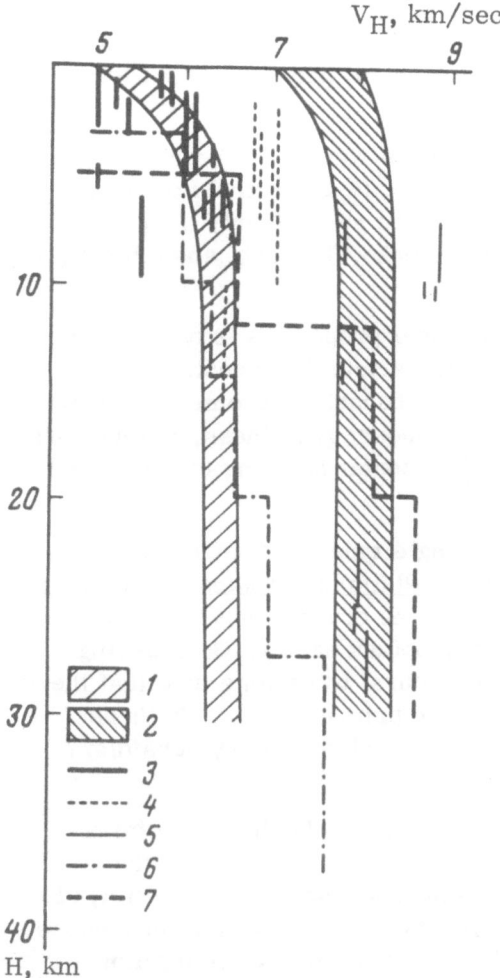

V_H, km/sec

Fig. 6.13. Comparison of the change in velocity with pressure for acidic and basic rocks with deep seismic sounding data: 1) rocks in the granite group; 2) gabbro, diorite; 3-5) boundary velocities for P_0^K, P_1^K, and P_0^M events in the transition zone between Asia and the Pacific Ocean [163]; 6) layer velocities for a continental crust and upper mantle; 7) the same, for an oceanic crust. The velocities of samples at various depths were taken from the curves on Fig. 6.10.

types which must be dominant at depth, from the petrological point of view. Work has already been partially completed in this direction for the Baltic Shield, in the Caucasus, and in Kazakhstan.

Inadequate regional data, finally, do not make it possible to develop general ideas about the possible composition of the crust and upper mantle in various regions, which would not be incompatible with the entire suite of contemporaneous geophysical and geological data.

Such an approach has been proposed in a new book by V. V. Belousov [19]. As several of the author's suggestions appear to us to be of a fundamental nature, I will review briefly his concepts of the nature of seismic layering in the earth's crust.

Based on an analysis of the distribution of thermal gradients and pressures in the crust and upper mantle in a variety of tectonic zones, as well as on ideas about the possible variations in the thermal regime and dynamic pressures during tectonogenic processes, V. V. Belousov proposes that in the crystalline part of a continental crust, during the formation of the crust over the depth range from 5 to 20 km, the conditions which prevailed were favorable for the development of granite or gneissic rocks. At moderate depths (15-30 km), most probably, in his opinion, the rocks belong to a granulite facies with gabbro intrusions, and he suggests that this layer be called the granulite-basalt layer. He further proposes that the M-discontinuity is a phase-change boundary representing the transition from basalt to eclogite, and he suggests that in the lowermost part of the crust, at a depth of 30-35 km, there is a layer that is a mixture of eclogite with the granulite facies characteristic of the middle part of the crust. He believes that the upper part of the mantle is of periodotite composition.

The Nature of Layering. Approximately horizontal deep-layering of the physical properties of complexly faulted rock masses, jumbled into folded and faulted sections with different blocks in intimate contact, appears to exist over major areas of ancient shields and stable platforms, and in intermontane and mountain foredeep basins of any age as a main feature of the seismic sections which have been discussed from various points of view in the preceding chapters. If we assume that the crust formed as a sequence of layers in ancient times in a number of areas, then it is obvious that in more recent years, the primary layering was frequently disrupted by tectonic processes so that the

crust in such areas has been reformed into the somewhat blocky structure which characterizes the changes in gross physical properties (primarily the increase of density and velocity
with depth) caused by the general increase in pressure, temperature, and degree of basic composition of the rock.

The structure of ancient shields and stable platforms, where rocks of different compositions with different physical properties are present at the surface, is an example of such a
blocky structure of the crust; the nearly horizontal layering can be seen clearly in spite of
these blocky and other structural macroinhomogeneities (Litvinenko, Sollogub, and others [115,
159-161, and others]).

It is very significant that the horizontal layering in physical properties is observed in
the study of the nature of the crust using seismic methods with various frequencies. The
higher the frequency is, the more complicated the seismic wave behavior becomes, reflecting
the structurally inhomogeneous character of the medium (blockiness, graininess, and nonhori-
zontal layering). However, the general tendency for nearly horizontal layering is preserved
(Litvinenko [115]).

A comparison of seismic sections for structures of a single type but different ages, as
for example, the Baltic shield and the Russian and Turkmenian Platforms, indicates that they
are quite similar. Similar layering is also found in the crystalline crust even when it is
buried in a basin (the south Caspian depression and the Predkopetdak basin). Considering
the differing histories of formation of these structures, it is natural to assume also that there
were significant differences in the processes of formation of the primary crust. In these
cases, we may assume that the seismic boundaries are reflections of secondary subsidiary
processes (Borisov [26-28], Fotiadi [178], Kosminskaya [217]).

The persistence of seismic boundaries within reasonably narrow depth ranges for a
single type of tectonic zone and the weak undulations of these boundaries, as well as the reflection of the primary features of the topography of the basement surface in their relief, allows us to assume that the development was related to the fronts of some physical processes
or the results of these processes. We have termed such fronts, fronts of metamorphism
(Riznochenko and Kosminskaya [149]). They may arise as a consequence of the combined effect
of changes in the pressure and temperature regime and changes in the material present at
depth in the crust and mantle. It may be that these fronts are associated with the boundaries
of zones of granitization, degranitization, and eclogitization of V. V. Belousov [19], or with
other similar features [168, 169, 181]. The positions of these zones may shift with depth in
relation to changes in the physical and chemical environment. When the processes come to an
end, the fronts stabilize and their movements cease. Belousov's [19] ideas on the limiting
positions of granitization fronts, and particularly of degranitization fronts, are of interest.
They may explain such facts as the essentially uniform depth to the seismic boundary with a
velocity of 6.5-6.7 km/sec on platforms and in basins, and the similarity of crustal sections
for shields and for ancient and young platforms.

Similar sections for these three types of structure may arise if, despite frequent differences in their "microstructures" and history of development, the main seismic boundaries
in the crust of such structures reach some critical depth, which is characterized by the
specific physical chemical transformation of a crust of average composition in a geosynclinal zone to a platform structure. It is interesting to note that a comparison of the ages
of these structures indicates that establishment of the critical depth apparently takes no more
time than the age of the youngest platform areas.

Subsequent reworking of these structures possibly results in a greater differentiation
between sections. We may evaluate this possibility by comparing the seismic sections for the

Turkmenian Platform (Ryaboi [152]) and the Predkopetdag basin (Vol'vovskii and others [42]), which is formed on top of a platform (see Fig. 6.8). The depth of the boundary with a velocity of 6.5 km/sec [42, 220] is somewhat less in the basin than on a platform, and the entire crustal section to a depth of about 30 km is characterized by high velocities. A characteristic feature of the lower part of the crust is the presence of a layer with a velocity of about 7.6 km/sec with a depth that is essentially the same as that for the M-discontinuity on the platform. The velocity in this zone is close to that for eclogite, but in Belousov's view, the formation of such a thick layer from the basaltic asthenolith is unlikely [19].

It is apparent that the assumption of an eclogitization process is not adequate to explain the formation of the boundary in the lower crust. It is obvious that in further work, we will have to develop other examples for the possible phase transitions of rocks under the conditions at these depths, pressures, and temperatures. However, even with such transitions, as are known, we may assume the existence of a phase related to the partial separation of some component which has a profound effect in changing the physical properties of the material at depth. This is possibly the course to pursue in the search for an explanation of velocity lamination of the upper mantle present on deep seismic sounding and seismological sections [117, 152].

Several geologists and geophysicists have advanced the concept of a crust-mantle mix, which is represented by a discretely structurally inhomogeneous grainy medium [117, 241]. Because of the discrete nature and properties of the components, the result may be a zone with relatively high or relatively low velocities in comparison with the layers above and below, respectively.

According to Belousov [19], it is important to recognize the depths at which "pure" granite may exist in a granitization zone. These may be less dense and be characterized by different elastic properties than the overlying slates and gneisses or the underlying rocks of the granulite facies. This is the basis for the hypothesized existence of relatively thin layers with low velocities in some areas.

Considering the complexity of the structure of the observed deep-traveling wave groups, the ragged boundaries between the zones of the various facies are of particular interest [19].

Deep seismic sounding data do not contradict Belousov's idea about the processes involved in thickening or thinning of the crust. In comparing the thick continental crust and the thin oceanic crust, we must assume that the thickening of the oceanic crust in the transitional region takes place either as a consequence of sediment accumulation, which when it sinks into the zone of granitization is subjected to metamorphism and is converted to a crystalline crust of the continental type, or as a result of aggradation from below, in relation with the phenomena of degranitization and eclogitization.

Thinning of the continental crust, as is indicated by the velocity section in transitional zones (see Fig. 6.4), is a result of spontaneous degradation of the lower part of the crust. In view of the fact that in areas where the M-discontinuity occurs at shallow depth, the velocity at this boundary is no more than 8 km/sec, we may hypothesize that the rise of this boundary from a depth of nearly 40 km in the northern part of the Magadan-Kolyma profile (Davydova, Yaroshevskaya, and Shvartz [63]) to 20 km in the central part of the Sea of Okhotsk is explained by a change in the level of the eclogitization level. The depths of rocks with velocities of 6.6-7.0 km/sec, which probably are related to a degranitization front, are practically invariant.

These examples, finally, do not prove Belousov's hypotheses [19], because similar geophysical data were used to some degree in formulating them. However, these examples are still useful because they were not used in the developments included in [19].

Despite its apparent validity, Belousov's hypothesis concerning the formation of the continental crust may not be a unique and complete explanation of all of the features of the seismic sections of the crust in various tectonic zones. In particular, the reason for the development of a thick layer with a velocity of 7.2-7.6 km/sec in the lower part of the crust, which would correspond to the velocity for eclogite, is still not clear. Aspects of the mechanism involved in layering of crustal and mantle materials not only in gross zones, but also in thinner "laminations" within these zones, are also unclear.

There is some disagreement between the velocities found in detailed deep seismic sounding sections and the values which were given in Belousov's book, which were averaged data for rocks of specific facies. In view of the fact that these differences fall within the velocity limits which have been observed for rocks in a single group, they may not be of any great value for judging the possible existence of various zones of metamorphism.

The main features of this hypothesis concerning the possibly multiply repeated shifts in the physical-chemical environment during the development of the crust from cycle to cycle under the action of a stress field (pressure) and temperature regime in the mantle and crust, concerning the relationship between these phenomena and the processes involved in asthenospheric waves, and the concept of the possible formation of a basaltic asthenolith and the injection of this from the mantle into the crust, are, in our opinion, an important step in the development of a theory for the geological interpretation of deep seismic sounding data.

Belousov's hypothesis allows more reliance to be placed on the geological feature developed earlier on the basis of the analysis of observed seismic-wave fields for a discontinuously layered or blocky-granular model of the crust.

An important feature of this hypothesis is the attempt to abandon the standard framework of a two-layer crust. However, this departure in [19] was not taken to lithologic completion. It would seem to us to be better to do away with the terms "granitic" and "basaltic" and use new terms which are more generally meaningful in terms of the actual composition of the crust, because various petrological facies which have properties that may be distinguished by geophysical means can be formed.

On the basis of our present knowledge of the nature of seismic boundaries, we may relate them to the following factors.

The surface of the crystalline part of the crust K_0 with a velocity of 6 km/sec corresponds to the surface of the metamorphosed and folded basement. In regions where there are deep basins, this boundary may correspond more precisely to a high velocity in the lower part of the sedimentary section. In mountainous regions, the concept of the surface of the crystalline crust is even less exact, and its surface may correspond simply to the surface of the folded and metamorphosed rocks.

The first boundary K_1 in the crystalline part of the crust, with a velocity of 6.3-6.5 km/sec, which is recognized in some areas at depths some 2 to 5 km greater than that of the surface of the crust K_0 probably corresponds to a change in the velocity gradient under the influence of pressure (see Fig. 6.11). The first boundary in the crystalline part of the crust in mountainous areas, in complexly folded metamorphic rocks is obviously a boundary of this type.

The deeper boundaries in the crystalline crust of the continents with velocities of 6.5-6.8 and 7.0-7.2 km/sec and the corresponding depths of 10-20 and 20-30 km may, apparently, be related to specific zones of intracrustal metamorphism, while boundaries in the lower crust at depths greater than 30 km and in the upper mantle (at depths up to 100 km and more) may be related to phase transitions (Subbotin, Magnitskii, Belousov, etc.).

In the case of the oceanic crust, the boundary between the crust and mantle apparently represents a marked change in the composition of the rocks [10, 11, 201].

The recognition of low-velocity layers with deep seismic sounding data presents a number of as yet, insurmountable difficulties. There are some deep seismic sounding data which seem to best fit with a crustal model that includes a thin (less than 5 km) low-velocity layer. However, at present, the characteristics of these layers can be determined only with low precision, no better than the limits of error of the method.

CONCLUSIONS

The questions which have been considered provide a basis for drawing conclusions about the effectiveness of the present-day deep seismic sounding method and some of the main courses of development, as well as indicating the relationship between deep seismic sounding and other methods.

§ 1.   Concerning the Deep Seismic Sounding Method

1. Deep seismic sounding is one of the methods of explosion seismology. It is based on the recording of deep-traveling waves from several shots, over the frequency range 3-20 Hz. Studies of the spectrum of the deep waves and the spectra of microseisms have indicated that the optimum frequency band in deep seismic sounding for operations both on land and at sea are determined by the resonant nature of the shot spectrum. Shot sizes of hundreds of kilograms of chemical explosives excite signals which at frequencies of 8-12 Hz on land and 4-6 Hz at sea are comparable with the microseismic noise level at large distances. The microseismic noise level varies according to the area (the minimum value at a frequency of 10 Hz is a tenth of an angstrom − 1×10^{-8} cm, and the maximum is 1×10^{-5} cm). These levels limit the sensitivity of the equipment that may be used, and so also, the recording distance, that is, the depth of reach of the method. Further significant improvement of the depth of penetration with deep seismic sounding for the same resolution will be possible with the introduction of new methods for extracting weak signals from the background noise either in the vicinity of the first refracted waves or in the vicinity of later reflected and refracted waves, superimposed on the background of stronger coherent and incoherent noise.

2. The resolution in deep seismic sounding is determined by the frequencies used and the spacing between receiver locations. The complicated nature of the seismic wave field observed on the continents makes it impractical to realize the advantages of high frequencies in recording individual waves, even with continuous profiling. Because of this, in deep seismic sounding, the schematic behavior of the seismic wave field is used by grouping specific waves into wave groups. The length of a group depends upon the detail with which the survey is made. The length is usually two or three times greater than the actual wavelength, which results in a corresponding reduction in the absolute resolution. With recording of refracted waves in a continental crust, layers with a thickness of 8-10 km may be recognized, while layers with a thickness of 2-5 km may be recognized when reflected waves are used. The maximum actual resolution of the deep seismic sounding method is realized with continuous profiling and the simultaneous use of travel-time curves for reflected and refracted waves, and is 4-7 km on the average.

Further improvement in the resolution of deep seismic sounding is possible in principle not by increasing the density of observations, which is already quite complicated, but rather by more complete interpretation of the records: the use of broadband recording of the wave field with subsequent frequency analysis and the recognition of new classes of waves (shear waves, transverse waves), more detailed studies in the area of high-angle reflections, inclusion of later waves of wave groups in quantitative interpretation, and so on.

3. At present, quantitative interpretation of deep seismic sounding is based on an analysis of the characteristics of the primary groups of deep-traveling waves. The sections constructed from these data are schematic and reflect only the average features of the wave field.

A comparison of the detailed structure of seismic wave fields with schematic representations in terms of wave groups has indicated that for a multiple layered crust, the use of group correlations leads to a number of systematic errors which cause the velocity in the crust to be underestimated when group correlations are used, with a corresponding decrease in the depths to reference horizons, including the M-discontinuity.

4. The methods of velocity and wave form interpretation in deep seismic sounding are based on models with continuous boundaries. These models do not reflect the important observed features of the seismic wave field — discontinuities in the correlation of individual waves. A discontinuous field may arise for several reasons: nonuniform boundaries — discontinuities in their properties, undulations, extent, and blockiness of the lower medium, as well as the phenomenon of interference of various wave forms related to a continuous gradational boundary.

Because of this, it is important to carry out investigations directed toward the development of a theory for seismic wave propagation in a complicated medium. Experimental studies in the field and on models must be used to provide values for the theoretical parameters in the corresponding models.

In the case of the basic theory for homogeneous and inhomogeneous layered models, it will be important to develop programs for computing seismic wave fields with consideration of the interference between all the possible waves which have an intensity within the dynamic range of the recording equipment.

5. Surveys run using various field techniques in differing tectonic zones reflect not only differences in structure but also the detail of the survey procedure.

With survey systems having differing degrees of detail, two types of seismic wave behavior may clearly be recognized — that for a continental crust and that for an oceanic crust. The main groups of waves are related to the surface of the crystalline part of the basement and to the M-discontinuity — the surface of the mantle. The characteristics of these surfaces provide a basis for comparing generalized deep seismic sounding sections of various types, and with seismological data.

6. The new ideas about the structure of the crust and upper mantle based on deep seismic sounding data differ significantly from earlier ideas. These indicate considerable velocity interlayering of the entire thickness of the crystalline crust and upper mantle; weak undulatory relief of deep blocks of crust, and rapid changes in the dip of these boundaries at the junction between two blocks; the nonuniform nature of the velocity characteristics of the M-discontinuity; and the existence of areas with a transitional boundary between the crust and mantle.

These data reflect the complexity of the processes involved in the formation of the crust, which must have a profound effect on the development of macroinhomogeneous blocks which exist along with the horizontal laminations of the deep materials with respect to velocity.

The basic question in deep seismic sounding is that of the further study of all possible details of the velocity section for the crust and mantle, including low-velocity layers, which would permit an elucidation of the nature of the deep layering and blocky macroinhomogeneities. Deep drilling results are essential in resolving this problem and for the further development of deep seismic sounding.

7. As a result of the present ideas of the complexity of the layering of the crust and upper mantle, it is quite clear not only that the earlier concept of a "granitic-basaltic" crust is not compatible with new data, but also that this outdated concept hampers the thinking of investigators in trying to generalize the narrow framework of a two-layer crust into that of a multiple layer crust formed by complicated processes.

§ 2. Relationship of Deep Seismic Sounding to Other Methods

Development of deep studies has involved the combined use of many geological and geophysical methods. Deep seismic sounding plays an important role in this approach. Deep seismic soundings provide the primary reference in interpreting many other methods.

In view of this, we should review the mutual interrelations between deep seismic sounding and other methods from the point of view that the results from several methods will prove more useful than the results of the methods interpreted separately.

1. Deep Seismic Sounding and Seismological Profiling. The development of novel techniques for studying the crust through the use of earthquakes as well as explosions is a logical extension of the ideas of G. A. Gamburtsev about the development of experimental seismology. The desirability for doing this in part is economic and in part a matter of expanding the capabilities of deep seismic sounding in terms of depth of investigation and of improving the resolution of seismological techniques for studying the crust and mantle.

The physical prerequisites for such a multiple-method approach will be a better understanding of the types of waves that are recorded, similarity of survey techniques and similarity of frequency ranges. There are difficulties involved in developing equipment which would permit recording signals from both small earthquakes and explosions with a broad range of energies with the same accuracy. These difficulties already have been partially overcome with the development of the "Zemlya" equipment — equipment for recording seismological data on magnetic tape, and which can be run unattended at remote locations for periods up to two weeks.

2. Deep Seismic Sounding and Gravimetry. With the study of the characteristics of the gravitational field and comparison with seismic sections, modern gravimetry displays a significant capability for evaluating deep crustal structure over large areas [7, 12, 51, 85, etc.]. However, in the solution of regional scale problems, it is clear that the correlation between a seismic section and the gravitational field varies widely and may deviate greatly from the statistical average for the earth in general or for its planetary structures, the continents, and oceans [28, 106]. These features are not always considered, resulting in some annoying misunderstandings and errors. Thus, gravity data are sometimes contoured with close-spaced contours, that is, with a precision of 1 km, while the deep seismic sounding data which provide the primary control for the gravity survey, reflect the same boundaries with an accuracy which is less by a factor of two to three. It is clear that such a formal accuracy is an illusion and may only result in leading nonspecialists astray.

We should note an important new direction which is being taken in the combined interpretation of seismic and other geophysical data. It is based on a very careful study of the correlation not only between crustal thickness and gravitational and magnetic anomalies, as has been done previously, but also with many parameters for the crust including factors other than geophysical (G. I. Karataev and others). Such a study is done in the same way in a methodical plan as in a regional plan: the comparison of specific situations. Such an approach is of primary interest for penetration into the deep mantle. Such features as the degree of magnetization (A. G. Gainanov, Yu. V. Tulina, and others), and possible increases or decreases in the density of various blocks — and recently within a single block — for various depths have been recognized. There are novel possibilities for the combined use of seismic data and temperature information, as well as data from deep magneto-telluric soundings.

We may expect that as a result of generalizing the geophysical data in various areas we will be able to reduce the amount of low-detail reconnaissance deep seismic sounding in studies of new areas, particularly in the oceans, by replacing this stage with a gravity net and occasional seismic profiles with a detail survey system.

3. Deep Seismic Sounding and the Geological Sciences. The ultimate objective in deep studies is not merely the determination of the structure of the earth, but the elucidation of those processes that have created it. The immediate results of deep seismic sounding specify only the present crustal structure. The extension to the processes involved in development will require a great deal of additional thought. Such an extension is particularly difficult if a single isolated area is considered. If deep seismic sounding data are obtained from several differing regions, comparison of the crustal structures for these regions and an association of crustal features with the geological history for these regions may lead to the possibility of determining the sequence of the processes. Both geology and geophysics are now widely used in this approach to the interpretation of deep seismic sounding. This leads to an evolutionary sequence for the crust — the direction of successive sections for various crustal structures from geosynclines to platforms, from continents to oceans (I. P. Kosminskaya, P. N. Kropotkin, V. V. Belousov, A. A. Borisov, and others).

As a result of the availability of new more-detailed and varied data on the crustal structure, there is a need for a new approach to interpreting the processes involved in its formation and evolution. In this scheme, creative contacts between the seismologists and geologists, physicists, and geochemists are particularly important, inasmuch as the geological (in the broad concept of the geological sciences) interpretation of deep seismic sounding has already gone beyond the framework of a geometrical concept of the depth to and relief of a boundary. In order to explain the nature of such boundaries, it will be necessary to study the processes related to the composition and behavior of deep materials under conditions of high pressure and temperature, for which the density of a rock may change not only by closing of pore structures and melting, but also by solid-state phase transitions.

REFERENCES

1. K. Aki and F. Press, "Structure of the upper mantle under the oceans and continents from Rayleigh-wave data," in: Verkhnaya Mantiya Zemli, Izd. Nauka (1964).
2. G. I. Aksenovich, E. I. Gal'perin, and M. A. Zaionchkovskii, "Characteristics of equipment of deep seismic investigations and results of its trials," Izv. Akad. Nauk SSSR, Ser. Geofizika, No. 2 (1957).
3. G. I. Aksenovich, L. E. Aronov, A. A. Gagel'gantz, E. I. Gal'perin, M. A. Zaionchkovskii, I. P. Kosminskaya, and R. M. Krakshina, Deep Seismic Sounding of the Earth's Crust in the Central Part of the Caspian Sea, Izd. AN SSSR (1962).
4. A. S. Alekseev, "On the kinematic and dynamic properties of deep waves in the cases of some theoretical models of the earth's crust," in: Glubinnoe seismicheskoe zondirovanie zemnoi kory v SSSR, Gostoptekhizdat (1962).
5. A. S. Alekseev, I. S. Vol'vovskii, N. I. Ermilova, P. V. Krauklis, and V. Z. Ryaboi, "The question of the physical nature of some waves recorded in deep seismic sounding, I-III," Izv. Akad. Nauk SSSR, Ser. Geofizika, No. 11 (1963); No. 1, 2 (1964).
6. A. S. Alekseev, Direct and Inverse Problems in Theoretical Seismology. Abstract, Doctoral Dissertation, Inst. Fiz. Zemli Akad. Nauk SSSR (1966).
7. B. A. Andreev, "Gravitational anomalies and the thickness of the earth's crust in continental areas," Dokl. Akad. Nauk SSSR, Vol. 119, No. 2 (1958).
8. A. N. Antonenko, Structure of the Earth's Crust in the Northern Part of Central Kazakhstan from Deep Seismic Sounding Data, Cand. Dissertation, Inst. Geol. Nauk, Akad. Nauk KazSSR, Alma-Ata (1964).
9. F. F. Aptikaev, "Parameters for seismic vibrations excited by explosions," in: Éksperimental'naya seismika, Tr. Inst. Fiz. Zemli Akad. Nauk SSSR, No. 32 (199) (1964).
10. G. D. Afanas'ev, "Structure of the earth's crust and some problems in petrology," Izv. Akad. Nauk SSSR, Ser. Geologiya, No. 3 (1961).
11. G. D. Afanas'ev, "More on the structure of the earth's crust from geophysical data and the position of petrography," Izv. Akad. Nauk SSSR, Ser. Geologiya, No. 10 (1962).
12. B. K. Balavadze, "Geophysical investigation of the structure of the earth's crust in the Black Sea Basin," in: Glubinnoe stroenie Kavkaza, Izd. Nauka (1966).
13. B. P. Belikov, Elastic Constants of Rockforming Minerals and Their Effect on the Elasticity of Rocks, Part 1, Izd. Nauka (1964).
14. V. V. Belousov, "Development of the earth and tectonogenesis," Sov. Geol., No. 7 (1960).
15. V. V. Belousov, "On the value of deep seismic sounding for the solution of theoretical and practical problems in geology," in: Glubinnoe seismicheskoe zondirovanie Zemnoi kory SSSR, Izd. Gostoptekhizdat (1962).
16. V. V. Belousov, "On the course of development of the geological sciences," Sov. Geol., No. 1 (1963).
17. V. V. Belousov, "On the crust and upper mantle of the continents," Sov. Geol., No. 1 (1965).

18.   V. V. Belousov, "Now the IGY and the prospects for further international cooperation,"
        Geophysical Bull. of the International Geophysical Committee, Akad. Nauk SSSR, No. 14,
        Izd. Nauka (1965).

19.   V. V. Belousov, The Earth's Crust and Upper Mantle of the Continents, Izd. Nauka (1965).

20.   V. V. Belousov, B. S. Vol'vovskii, I. S. Vol'vovskii, and V. Z. Ryaboi, "Experimental
        investigation of the recording of deep reflected waves," Izv. Akad. Nauk SSSR, Ser.
        Geofizika, No. 8 (1962).

21.   N. A. Belyaevskii and V. V. Fedinskii, "Studies of the deep earth and problems in super-
        deep drilling," Sov. Geol., No. 12 (1961).

22.   I. S. Berzon, "On the resolution of the seismic methods," Izv. Akad. Nauk SSSR, Ser.
        Geograf. i Geofiz., No. 3 (1947).

23.   I. S. Berzon, High-Frequency Seismic Exploration, Izd. AN SSSR (1957).

24.   I. S. Berzon, A. M. Epinat'eva, G. N. Pariiskaya, and S. P. Starodubrovskaya, Dynamic
        Characteristics of Seismic Waves in Real Media, Izd. Akad. Nauk SSSR (1962).

25.   I. S. Berzon, "On the development of the physical basis of seismic methods of explora-
        tion," Izv. Akad. Nauk SSSR, Ser. Geofiz., No. 5 (1964).

26.   A. A. Borisov, "On the evolution of the earth's crust and the process of tectonogenesis,"
        Izv. Akad. Nauk SSSR, Ser. Geofiz., No. 2 (1963).

27.   A. A. Borisov, "Morphology of the Mohorovičić surface in the USSR and its structural
        significance," Sov. Geol., No. 4 (1964).

28.   A. A. Borisov, Deep Structure in the USSR from Geophysical Data. Abstract, Doct.
        Dissertation, Inst. Geol. Nauk, Akad. Nauk SSSR (1965).

29.   N. K. Bulin and Yu. I. Sitin, "Use of seismic data for studying the deep structure of
        the earth's crust in Turkmenia," in: Problema neftegazonosnosti Srednei Azii, No. 42,
        Gostoptekhizdat (1960).

30.   V. I. Bune', Attempt at a Detailed Analysis of the Seismicity of the Bakhshsk Region of
        the Tadzhik SSR, Doctoral Dissertation, Inst. Fiz. Zemli Akad. Nauk SSSR (1966).

31.   E. M. Butovskaya and V. I. Ulomov, "Seismic travel-time curves and some features of
        the crustal structure in Central Asia from data recorded from large explosions,"
        in: Glubinnoe seismicheskoe zondirovanie zemnoi kory v SSSR, Gostoptekhizdat (1962).

32.   E. M. Butovskaya, A. I. Zakharova, Kh. A. Atabaev, and Yu. P. Flenov, "Results of the
        use of specific travel-time curves for determining the epicenters in some regions of
        Central Asia," in: Izuchenie vnutrennego stroeniya zemli po seismicheskim dannym,
        Bull. Soveta po Seismologii, No. 15, Izd. Akad. Nauk SSSR (1963).

33.   V. G. Vasil'ev, P.S. Veitsman, E. I. Gal'perin, V. A. Gladun, A. V. Goryachev, S. M.
        Zverev, I. P. Kosminskaya, et al., "Investigation of the earth's crust in the region of
        transition from the Asiatic continent to the Pacific Ocean in 1958," in: Seismicheskie
        issledovaniya, No. 4, Izd. Akad. Nauk SSSR (1960).

34.   P. S. Veitsman, "On realizing the principle of reciprocity in seismology," Izv. Akad.
        Nauk SSSR, Ser. Geogr. i Geofiz., No. 3 (1948).

35.   P. S. Veitsman, "On the selection of spacings between seismographs for patterns for
        the purpose of decreasing the noise background," Izv. Akad. Nauk SSSR, Ser. Geofiz.,
        No. 6 (1952).

36.   P. S. Veitsman, "Correlation of seismic waves for deep seismic sounding of the
        earth's crust," Izv. Akad. Nauk SSSR, Ser. Geofiz., No. 12 (1957).

37.   P. S. Veitsman, E. I. Gal'perin, I. P. Kosminskaya, B. S. Vol'vovskii, Yu. N. Godin,
        N. P. Ivanova, and I. V. Pomerantseva, "Method of deep seismic sounding on land and
        at sea," in: Glubinnoe seismicheskoe zondirovanie zemnoi kory v SSSR, Gostoptekhizdat
        (1962).

38.   P. S. Veitsman and I. P. Kosminskaya, "Regional special seismological investigations
        during the IGY," in: Seismologicheskie issledovaniya, No. 5, Izd. Akad. Nauk SSSR (1964).

39.   P. S. Veitsman, "Features of the deep structure of the Kuriles-Kamchatka zone," Izv. Akad. Nauk SSSR, Fiz. Zemli, No. 9 (1965).

40.   M. P. Volarovich, "Investigation of the elastic and attenuation properties of rocks under high pressure," in: Tr. VI Soveshchaniya po Eksper. i Tekhn. Mineralogii (1962).

41.   M. P. Volarovich, N. E. Galdin, and A. I. Levikin, "Investigation of the velocity of longitudinal waves in samples of igneous and metamorphic rocks at pressures to 20,000 kg/cm^2," Izv. Akad. Nauk SSSR, Fiz. Zemli, No. 3 (1966).

42.   I. S. Vol'vovskii, Seismic Investigation of the Earth's Crust in Western Central Asia, Cand. Dissertation, Inst. Fiz. Zemli Akad. Nauk SSSR (1963).

43.   G. I. Petrashen' (editor), Aspects of the Dynamic Theory of Propagation of Seismic Waves, I, 1957; II, III, 1959, Izd. Leningr. Gos. Univ.

44.   I. N. Galkin and N. N. Kichin, "Use of model SS 30/60 seismic amplifier equipment for deep seismic sounding," Razv. i Promysh. Geofiz., No. 42, Gostoptekhizdat (1961).

45.   I. N. Galkin and M. A. Zaionchkovskii, "On equipment for deep seismic sounding," in: Glubinnoe seismicheskoe zondirovanie zemnoi kory v SSSR, Gostoptekhizdat (1962).

46.   I. N. Galkin, "Use of the absolute amplitude features of waves for aligning equipment for operations with deep seismic sounding at sea," Izv. Akad. Nauk SSSR, Fiz. Zemli, No. 11 (1966).

47.   E. I. Gal'perin and I. P. Kosminskaya, "Features of the deep seismic sounding method at sea," Izv. Akad. Nauk SSSR, Ser. Geofiz., No. 7 (1958).

48.   G. A. Gamburtsev, Yu. V. Riznochenko, I. S. Berzon, A. M. Epinat'eva, I. P. Pasechnik, I. P. Kosminskaya, and E. V. Karus, Correlation of Refracted Waves, Izd. Akad. Nauk SSSR (1952).

49.   G. A. Gamburtsev, Basis of Seismic Exploration, 2nd ed., Gostoptekhizdat (1959).

50.   G. A. Gamburtsev, Selected Works, Izd. Akad. Nauk SSSR (1960).

51.   A. G. Gainonov, Yu. V. Tulina, I. P. Kosminskaya, S. M. Zverev, P. S. Veitsman, and O. N. Solov'ev, "Combined interpretation of geophysical data in the Sea of Okhotsk and the Kuriles-Kamchatka zone of the Pacific Ocean," in: Seismicheskie issledovaniya, Izd. Nauka (1965).

52.   Yu. N. Godin, B. S. Vol'vovskii, and I. S. Vol'vovskii, "Seismic investigations of the earth's crust in the Bukhara region of the Uzbek SSR," Dokl. Akad. Nauk SSSR, Vol. 134, No. 5 (1960).

53.   Yu. N. Godin, B. S. Vol'vovskii, and I. S. Vol'vovskii, "Seismic investigation of the earth's crust in the region of the Fergana intermontane basin," Dokl. Akad. Nauk SSSR, Vol. 133, No. 6 (1960).

54.   Yu. N. Godin and A. V. Egorkin, "Structure of the earth's crust from data from a regional seismic survey in the southeastern Russian platform," Dokl. Akad. Nauk SSSR, Vol. 135, No. 5 (1960).

55.   Yu. N. Godin, B. S. Vol'vovskii, I. S. Vol'vovskii, and K. E. Fomenko, "Study of crustal structure in regional seismic investigations on the Russian platform and in Central Asia," Izv. Akad. Nauk SSSR, Ser. Geofiz., No. 10 (1961).

56.   Yu. N. Grachev, M. Ya. Dekhnich, I. V. Litvinenko, K. A. Nekrasova, and A. V. Sosnovskaya, "Deep geophysical investigation in the Baltic shield," Proc. of the 21st International Geological Congress, 1959, in: Geologicheskie rezul'taty prikladnoi geokhimii i geofiziki, Ch. 2, geofizika, Gostoptekhizdat (1960).

57.   F. Gron, "Reflection of a spherical impulsive wave from a plane interface," Tr. Geofiz. Inst. Czech. Akad. Nauk, No. 151 (1961).

58.   G. Z. Gurarii and I. A. Solov'eva, "Crustal structure from geophysical data," Tr. Geol. Inst. Akad. Nauk SSSR, Vol. 98, Izd. Akad. Nauk SSSR (1963).

59.   I. I. Gurvich, Seismic Exploration, Gostoptekhizdat (1960).

60.   I. I. Gurvich, "Theory of seismic radiation of spherical waves," Izv. Akad. Nauk SSSR, Fiz. Zemli, No. 10 (1965).

61.   I. I. Gurvich, V. B. Levyant, and L. V. Molotova, "Experimental amplitude spectra of explosions," Izv. Akad. Nauk SSSR, Fiz. Zemli, No. 3 (1966).

62.   Beno Gutenberg, Physics of the Earth's Interior.

63.   N. I. Davydova, Ya. B. Shvartz, and G. A. Yaroshevskaya, "Wave maps for deep seismic investigations on the profile Magadan-Kolima," in: Glubinnoe seismicheskoe zondireovanie zemnoi kory v SSSR, Gostoptekhizdat (1962).

64.   N. I. Davydova, G. V. Krasnopevtzeva, S. A. Manilov, V. A. Levi, L. A. Lobastova, E. M. Shekinskii, and G. K. Tvaltvadze, "Results of deep seismic sounding in the Caucasus," in: Glubinnoe stroenie Kavkaza, Izd. Nauka (1966).

65.   R. M. Demenitzkaya, "Basic features of crustal structure from geophysical data," Tr. Nauchn.-issled. Inst. Geologii Arktiki, Vol. 115, Gostoptekhizdat (1961).

66.   Yu. B. Demidenko, M. G. Manyuta, V. A. Lisenko, and L. M. Spikhina, Results of Seismic Investigations of the Deep Structure of the Earth's Crust in the Eastern Ukraine, Geofiz. Sb., Inst. Geofiz., Akad. Nauk Ukr. SSR, Iss. 5 (7) (1963).

67.   H. Jeffreys, Earth, Cambridge Univ. Press, 4th ed. (1960).

68.   A. T. Donabedov, T. A. Korovina, and K. V. Timarev, "Study of the crustal structure on the eastern shore of the Caspian Sea with the deep seismic sounding method," in: Glubinnoe seismicheskoe zondirovanie zemnoi kory v SSSR, Gostoptekhizdat (1962).

69.   A. V. Egorkin, Aspects of the Method of Determining the Characteristics of the Earth's Crust, Cand. Dissertation, Inst. Fiz. Zemli Akad. Nauk SSSR (1965).

70.   A. M. Epinat'eva, "On reflected waves arising with angles of incidence greater than the limiting value," Izv. Akad. Nauk SSSR, Ser. Geofiz., No. 6 (1957).

71.   A. M. Epinat'eva, Study of Longitudinal Seismic Waves Propagating in Some Real Layered Media, Tr. Inst. Fiz. Zemli Akad. Nauk SSSR, No. 14 (181) (1960).

72.   A. M. Epinat'eva, "The intensity of multiply reflected waves," in: Seismicheskie mnogokratnie otrazhennie volny. Tr. Inst. Fiz. Zemli, Akad. Nauk SSSR, Vol. 201, No. 34 (1964).

73.   A. M. Epinat'eva and V. F. Chervini, "Reflected waves in the vicinity of the second initial point," Studia Geoph. et Geod., No. 9 (1965).

74.   A. M. Epinat'eva and E. V. Karus, "Head waves from thin layers," in: Modeli real'-nykh sred i seismicheskie volnovie polya, Izd. Nauka (1967) [Seismic Wave Propagation in Real Media, Consultants Bureau, New York (1969)].

75.   V. N. Zharkov, "On the physical cause of waveguides in the outer shells of the earth at depths of 50-200 km," Dokl. Akad. Nauk SSSR, Vol. 125, No. 4 (1959).

76.   M. A. Zaionchkovskii, "On the absolute magnitudes of seismic signals observed in deep seismic sounding," in: Glubinnoe seismicheskoe zondirovanie zemnoi kory v SSSR, Gostoptekhizdat (1962).

77.   S. M. Zverev, "On the multiplicity of refracted and reflected waves in the ocean," in: Glubinnoe seismicheskoe zondirovanie zemnoi kory v SSSR, Gostoptekhizdat (1962).

78.   S. M. Zverev, "Frequency characteristics for shots in deep seismic sounding in the deep ocean," Izv. Akad. Nauk SSSR, Ser. Geofiz., No. 3 (1963).

79.   S. M. Zverev, Seismic Investigations at Sea, Izd. Moskovsk. Gos. Univ. (1964).

80.   S. M. Zverev, "Methods of studying the sedimentary section in deep seismic sounding at sea," in: Voprosy metodiki glubinnogo seismicheskogo zondirovaniya, Izd. Nauka (1965).

81.   S. M. Zverev and Yu. V. Tulina, "Some results of a detailed study of crustal structure in the southern part of the Kurile Islands with the deep seismic sounding method," in: Vulkanizm i glubinnoe stroenie zemli, Izd. Nauka (1966).

82.   S. M. Zverev and I. N. Galkin, "Survey procedures and the feasibility of improving recording in deep seismic soundings at sea," Izv. Akad. Nauk SSSR, Fiz. Zemli, No. 9 (1966).

83. T. G. Ivanova and Yu. I. Vasil'ev, "On the choice of the optimum characteristics for equipment for recording head waves from the crystalline basement," Izv. Akad. Nauk SSSR, Ser. Geofiz., No. 5 (1964).

84. D. N. Kazanli and A. A. Popov, "The characteristics of deep waves recorded in the central Caucasus," in: Glubinnoe seismicheskoe zondirovanie zemnoi kory v SSSR, Gostoptekhizdat (1962).

85. G. I. Karataev, Linear Predictions: The Structure and the Composition of the Earth's Crust According to Gravitational and Magnetic Data (Abstract of Doctoral Dissertation) (1964).

86. V. M. Kovilin and Yu. P. Neprochnov, "Structure of the earth's crust and the sedimentary sequence in the central part of the Sea of Japan according to seismic data," Izv. Akad. Nauk SSSR, Ser. Geol., 4 (1965).

87. I. K. Kondrat'ev, "Investigation of the frequency characteristics of some inhomogeneous layers," Izv. Akad. Nauk SSSR, Fiz. Zemli, No. 8 (1965).

88. O. K. Kondrat'ev and A. G. Gamburtsev, Seismic Investigation of the Near-Shore Part of Eastern Antarctica, Izd. Akad. Nauk SSSR (1963).

89. I. P. Kosminskaya, "On the initial points of travel-time curves for Mintrop waves," Izv. Akad. Nauk SSSR, Ser. Geogr. i Geofiz., No. 1 (1946).

90. I. P. Kosminskaya, "Interference of seismic waves excited by a harmonic source," Izv. Akad. Nauk SSSR, Ser. Geofiz., No. 4 (1952).

91. I. P. Kosminskaya, "Amplitude curves and phase plots of seismic waves generated by a harmonic point force in a homogeneous ideally elastic halfspace," Tr. Geofiz. Inst. Akad. Nauk SSSR, No. 30 (157) (1955).

92. I. P. Kosminskaya, "Methods of analyzing amplitude curves and phase plots for complex harmonic waves," Tr. Geofiz. Inst. Akad. Nauk SSSR, No. 30 (157) (1955).

93. I. P. Kosminskaya, "Analysis of the interference zone for seismic waves," Tr. Geofiz. Inst. Akad. Nauk SSSR, No. 35 (162) (1956).

94. I. P. Kosminskaya, "On the use of deep seismic sounding in various parts of the USSR," Byull. Soveta po Seismologii Akad. Nauk SSSR, No. 3 (1957).

95. I. P. Kosminskaya and Yu. V. Tulina, "Attempt at the use of the deep seismic sounding method for studying crustal structure in some areas of southwestern Turkmenia," Izv. Akad. Nauk SSSR, Ser. Geofiz., No. 7 (1957).

96. I. P. Kosminskaya, G. G. Mikhota, and Yu. V. Tulina, "Crustal structure in the Pamir—Alai zone from deep seismic sounding data," Izv. Akad. Nauk SSSR, Ser. Geofiz., No. 10 (1958).

97. I. P. Kosminskaya, "Crustal structure from seismic data," Byull. Moskovsk. Obshchestva Ispytat prirody, Otd. Geol., No. 4 (1958).

98. I. P. Kosminska and R. M. Krakshina, "On near-critical reflections from the Mohorovičic boundary," Izv. Akad. Nauk SSSR, Ser. Geofiz., No. 6 (1961).

99. I. P. Kosminskaya, "Crustal structure in the deep-water basins of the Black, Caspian, Japan, Okhotsk, and Bering Seas," Byull. Moskovsk. Obshchestva Ispytat. prirody, Otd. Geol., No. 6 (1961).

100. I. P. Kosminskaya, "On the relationship between high-detail and low-detail surveys," in: Glubinnoe seismicheskoe zondirovanie zemnoi kory v SSSR, Gostoptekhizdat (1962).

101. I. P. Kosminskaya, "Classification of crustal structure from seismic data," Byull. Soveta po Seismologii, No. 10 (1963).

102. I. P. Kosminskaya, S. M. Zverev, P. S. Veitsman, Yu. V. Tulina, and R. M. Krakshina, "Basic features of crustal structure in the Sea of Okhotsk and the Kuriles-Kamchatka zone of the Pacific Ocean from deep seismic sounding data," Izv. Akad. Nauk SSSR, Ser. Geofiz., No. 10 (1963).

103.   I. P. Kosminskaya, "Crustal studies in the USSR during the IGY," in: Rezul'taty iss-
        ledovanii po programme MGG, Seismologicheskoe issledovaniya, No. 5, Akad. Nauk
        SSSR (1963).

104.   I. P. Kosminskaya, Crustal Studies in the USSR during the IGY, Geofiz. Byull.
        Mezhduvedomstvennogo Geofiz. Komiteta, No. 14, Nauka (1965).

105.   I. P. Kosminskaya, "Aspects of deep seismic sounding of the earth's crust at the XII
        Assembly of the IGY Committees," Izv. Akad. Nauk SSSR, Fiz. Zemli, No. 5 (1965).

106.   I. P. Kosminskaya and Yu. M. Sheinmann, Features of Crustal Structure in Mountain
        Foredeeps and Intermontane Basins, Byull. Moskovsk. Obshchestva Ispytat. prirody,
        No. 4 (1965).

107.   I. P. Kosminskaya, in: Development of the Deep Seismic Sounding Method (a volume
        of papers commemorating the 60th birthday of G. A. Gamburtsev), Nauka (1966).

108.   I. P. Kosminskaya, in: Present-Day Problems in Deep Seismic Sounding (a volume of
        papers commemorating the 60th birthday of S. I. Subbotin), Naukova Dumka (1966).

109.   I. P. Kosminskaya and S. M. Zverev, "Problems in seismic investigations in the tran-
        sition zone between continents and oceans," in: Stroenie i razvitie zemnoi kory na
        Sovetskom Dal'nem Vostoke, Nauka (1968).

110.   V. N. Krestnikov and I. L. Nersesov, "Tectonic structure of the Pamirs and Tien-
        Shan and its relation to relief on the Mohorovičic," Sov. Geol., No. 11 (1962).

111.   N. V. Kuz'mina, A. N. Romashev, V. G. Rulev, D. A. Kharin, and E. I. Shemyakina,
        Seismic Effects of Explosions in the Scatter of Related Groups, Aspects of En-
        gineering Seismology, Tr. Inst. Fiz. Zemli, Akad. Nauk SSSR, No. 21 (188) (1962).

112.   I. S. Lenina, "Refracted waves and the study of seismic sections in the Pechangsk
        Region," Zap. Leningr. Gornogo Inst., Vol. 66, No. 2 (1963).

113.   I. V. Litvinenko, M. Ya. Dekhnich, and K. A. Nekrasova, "Deep seismic sounding in the
        Region of the Baltic Shield, Seismolog. Issledov. Sb. statei, No. 4, Izd. Akad. Nauk
        SSSR (1960).

114.   I. V. Litvinenko and K. A. Nekrasova, "Features of deep seismic sounding on the
        Baltic Shield," in: Glubinnoe seismicheskoe zondirovanie zemnoi kory v SSSR, Gos-
        toptekhizdat (1962).

115.   I. V. Litvinenko, "New seismic data on the crustal structure of the Baltic shield,"
        Dokl. Akad. Nauk SSSR, Vol. 149, No. 6 (1963).

116.   I. V. Litvinenko, "The seismic method for studying the deep structure of the Baltic
        shield," Zap. Leningr. Gornogo Inst., Vol. 66, No. 2 (1963).

117.   V. A. Magnitskii, Internal Structure and Physics of the Earth, Nedra (1965).

118.   S. I. Masarskii, "Travel-time curves for seismic waves in the Alai for records ob-
        tained from industrial explosions," Izv. Akad. Nauk SSSR, Ser. Geofiz., No. 7 (1962).

119.   Yu. V. Riznochenko (Editor), "Methods for detailed study of seismicity," Tr. Inst.
        Fiz. Zemli Akad. Nauk SSSR, No. 9 (176) (1960).

120.   N. G. Mikhailova, B. S. Pariiskii, and M. V. Saks, "Spectral characteristics of groups
        of layers," Izv. Akad. Nauk SSSR, Fiz. Zemli, No. 1 (1966).

121.   G. G. Mikhota, "Results of some attempts at studying the relationship of intensity
        and frequency spectra to charge weight," in: Glubinnoe seismicheskoe zon-
        dirovanie zemnoi kory v SSSR, Gostoptekhizdat (1962).

122.   G. G. Mikhota, "The dependence of the intensity and the frequency spectrum of
        waves on charge weight," in: Voprosy metodiki glubinnogo seismicheskoe zondiro-
        vaniya, Nauka (1965) [Problems in Deep Seismic Sounding, Consultants Bureau,
        New York (1967)].

123.   G. G. Mikhota and Yu. V. Tulina, "Experiments with grouped shot holes for deep
        seismic sounding," in: Voprosy metodiki glubinnogo seismicheskoe zondirovaniya,
        Nauka (1965) [Problems in Deep Seismic Sounding, Consultants Bureau, New York
        (1967)].

124.   Yu. P. Neprochnov, "Results of deep seismic sounding in the Black Sea," in: Glubinnoe seismicheskoe zondirovanie zemnoi kory v SSSR, Gostoptekhizdat (1962).

125.   Yu. P. Neprochnov, V. M. Kovilin, V. V. Zdorovenin, and V. Ya. Karp, "New data on crustal structure in the Japan Sea," Dokl. Akad. Nauk SSSR, Vol. 155, No. 6 (1964).

126.   A. A. Lukk and I. L. Nersesov, "Structure of the mantle from earthquake records," Dokl. Akad. Nauk SSSR, Ser. Matem. Fiz., Vol. 162, No. 3 (1965).

127.   I. L. Nersesov and T. G. Rautian, "Kinematics and dynamics of seismic waves at distances up to 3500 km from the epicenter," in: Eksperimental'naya seismika, Tr. Inst. Fiz. Zemli Akad. Nauk SSSR, No. 32 (199) (1964).

128.   I. L. Nersesov and A. V. Nikolaev, On the Dependence of Dominant Frequencies of Explosions to Charge Size, Tr. Inst. Fiz. Zemli Akad. Nauk SSSR, No. 25 (192)(1962).

129.   I. V. Pomerantseva and M. V. Margot'eva, Concerning the Question of the Origin of Waves Recorded in Deep Seismic Sounding, Izd. VNIIGeofizika (1959).

130.   I. V. Pomerantseva and M. V. Margot'eva, "Concerning the question of the origin of waves recorded in deep seismic sounding," in: Glubinnoe seismicheskoe zondirovanie zemnoi kory v SSSR, Gostoptekhizdat (1962).

131.   I. V. Pomerantseva Origin of Waves and Features of the Interpretation of Refraction and Deep Seismic Sounding Data Obtained in the Study of Crustal Structure in the Southeastern Russian Platform (Abstract, Cand. Dissertation), Inst. Fiz. Zemli (1964).

132.   I. V. Pomerantseva et al., "Use of the 'Zemlya' seismic equipment for studying structure in the southeastern Russian platform," Dokl. Akad. Nauk SSSR, Vol. 163, No. 1 (1965).

133.   M. K. Polshkov, Basic Aspects of Seismic Instrumentation, Gostoptekhizdat (1962).

134.   A. A. Popov, Crustal Structure in the Central Caucasus According to Deep Seismic Sounding Data, Cand. Dissertation, Inst. Geol. Nauk, Akad. Nauk KazSSR (1964).

135.   N. N. Puzyrev, Interpretation of Seismic Exploration Data with the Reflected-Wave Method, Gostoptekhizdat (1959).

136.   N. N. Puzyrev, "On interpretation of data from the refraction method in terms of the velocity gradient in the lower medium," Geologiya i Geofizika, No. 10, Izd. Akad. Nauk SSSR (1960).

137.   N. N. Puzyrev, "Concerning the question of the use of simplified surveying techniques for studying the basement in the West Siberian downwarp with the reflected-wave method," Geologiya i Geofizika, Izd. SO AN SSSR, No. 11 (1960).

138.   N. N. Puzyrev, V. A. Kondrashov, S. V. Krylov, and S. V. Potap'ev, "First results of deep seismic sounding studies of crustal structure in the central part of Western Siberia," Geologiya i Geofizika, SO AN SSSR, Izd., No. 11 (1964).

139.   N. N. Puzyrev, S. V. Krylov, and S. V. Potap'ev, "Point seismic sounding," in: Metodika seismorasvedki, Nauka (1965).

140.   M. M. Radzhabov and O. B. Baba-Zade, "On reflected-diffracted waves recorded in deep seismic sounding of the earth's crust," Izv. Akad. Nauk SSSR, Fiz. Zemli, No. 5 (1966).

141.   T. G. Rautian, "On determining the energy of earthquakes at distances up to 3000 km," in: Éksperimental'naya seismika, Tr. Inst. Fiz. Zemli Akad. Nauk SSSR, No. 32 (199) (1964).

142.   I. A. Rezanov, "On the structure of the earth's crust in platform areas," Byull. Mosk. Ob. Ispit. Prirodi, Otd. Geol., No. 1 (1962).

143.   R. Raitt, "Studies of the Pacific Ocean basin with refraction seismic methods," in: Structure of the Earth's Crust from Seismic Data [Russian translation] (1959).

144.   Yu. V. Riznochenko, "Seismic velocity in layered media," Izv. Akad. Nauk SSSR, Ser. Geogr. i Geofiz., No. 2 (1947).

145.   Yu. V. Riznochenko, "On the determination of effective velocity under the condition of poorly correlated reflections," Izv. Akad. Nauk SSSR, Ser. Geogr. i Geofiz., No.1 (1948).

146. Yu. V. Riznochenko, "On the study of crustal structure during the IGY," Izv. Akad. Nauk SSSR, Ser. Geofiz., No. 2 (1957).

147. Yu. V. Riznochenko, "On deep seismic sounding work in the USSR," Studia Geofiz. et Geod. (1958).

148. Yu. V. Riznochenko, "Reference earth profiles," Vestn. Akad. Nauk SSSR, No. 8 (1961).

149. Yu. V. Riznochenko and I. P. Kosminskaya, "On the origin of layering in the earth's crust and upper mantle," Dokl. Akad. Nauk SSSR, Vol. 155, No. 2 (1963).

150. A. E. Ringwood, "Models of the upper mantle," in: Verkhnyaya mantiya Zemli, Izd. Mir (1964).

151. G. B. Rybak and S. P. Ol'shtinskii, "Concerning the question of the relationship of amplitude and frequency spectrum of longitudinal waves to charge weight," in: Geol. interpretatsiya i metodika geofiz. issledovanii, Izd. Naukova Dumka (1964).

152. V. Z. Ryaboi, "Structure of the earth's crust and upper mantle along the deep seismic sounding profile Kopetdag – Aral Sea," Sov. Geol., No. 5 (1963).

153. V. Z. Ryaboi, Study of the Character of Deep Waves in Seismic Studies of the Earth's Crust and Upper Mantle with the Deep Seismic Sounding Method (Abstract of Cand. Dissertation), Inst. Fiz. Zemli Akad. Nauk SSSR (1965).

154. V. Z. Ryaboi, "Measurement of the absolute amplitude characteristics of waves in deep seismic sounding and refraction shooting," Izv. Akad. Nauk SSSR, Fiz. Zemli, No. 6 (1965).

155. V. Z. Ryaboi, "Kinematic and dynamic characteristics of waves related to boundaries in the earth's crust and upper mantle," Izv. Akad. Nauk SSSR, Fiz. Zemli, No. 3 (1966).

156. E. F. Savarenskii, "Study of the outer shells of the earth with seismic body waves," Izv. Akad. Nauk SSSR, Fiz. Zemli, No. 3 (1966).

157. N. S. Smirnova and N. I. Ermilova, "On the construction of theoretical seismograms in the initial point," in: Voprosy dinamicheskoi teorii rasprostraneniya seismicheskikh voln, Vol. III, Izd. Leningr. Gos. Univ. (1959).

158. N. S. Smirnova, "Computation of wave fields in the vicinity of principal points," in: Voprosy dinamicheskoi teorii rasprostraneniya seismicheskikh voln, No. 6, Izd. Leningr. Gos. Univ. (1962).

159. V. B. Sollogub, A. V. Chekunov, L. T. Kalyushnaya, and A. A. Khilinskii, "On the deep structure of the Korosten pluton from seismic data," Dokl. Akad. Nauk SSSR, Vol. 152, No. 5 (1963).

160. V. B. Sollogub, A. V. Chekunov, and N. I. Pavlenkova, "Results of deep seismic sounding along the profile Black Sea – Voronezh Massif," in: Proceedings of the VII Congress of the Carpato-Balkan Geological Association, Part 7, Sofia (1965).

161. V. B. Sollogub, A. V. Chekunov, N. I. Pavlenkova, et al., "Some features of the wave picture in zones of fracture of the earth's crust in the Ukrainian SSR," in: Stroenie i razvitie zemnoi kory, No. 1 (12) (1965).

162. P. A. Storev and A. G. Gainanov, "On the structure of the earth's crust in the Indian Ocean from data of geophysical investigations," Okeanologiya, Vol. V, No. 4, Nauka (1965).

163. E. I. Gal'perin and I. P. Kosminskaya (Editors), Structure of the Earth's Crust in the Transition Region from the Asian Continent to the Pacific Ocean, Nauka (1964).

164. S. I. Subbotin, V. B. Sollogub, and A. V. Chekunov, "Crustal structure of the basic geostructural elements in the Ukraine," Dokl. Akad. Nauk SSSR, Vol. 153, No. 2 (1963).

165. S. I. Subbotin, G. L. Naumchik, and I. Sh. Rakhimova, Processes in the Upper Mantle and Their Relation to the Structure of the Earth's Crust, Naukova Dumka (1964).

166. E. D. Tagai, "Use of the Vikherta-Chibisova method for determining the velocity parameters of media in deep seismic sounding," Prikladn. Geofizika, No. 36 (1960).

167. G. K. Tvaltvadze, Crustal Structure in Georgia from Seismic Data and Construction of a System of Theoretical Travel-Time Curves, Izd. AN GruzSSR (1960).

168.   V. V. Tikhomirov, "On the question of the development of the earth's crust and con-
       cerning the significance in this process of the phenomenon of metasomatosis," in:
       Proc. of the 21st International Geological Congress, Kiev (1960).

169.   V. V. Tikhomirov, "On the relationships of alteration processes in the earth's crust
       with a vertical direction of tectonic movement," Dokl. Akad. Nauk SSSR, Vol. 151,
       No. 5 (1963).

170.   Yu. V. Tulina, "Comparison of magnetic anomalies with the seismic properties of the
       Mohorovičic boundary," Izv. Akad. Nauk SSSR, Fiz. Zemli, No. 3 (1965).

171.   G. B. Udintsev, Relief of the Ocean Bottom and Aspects of Tectonics. Marine Geol-
       ogy and Sea Dynamics, Izd. AN SSSR (1962).

172.   G. B. Udintsev, "New map of the relief of the Pacific Ocean bottom," Okeanologiya,
       Vol. 3, No. 1 (1963).

173.   N. E. Fedoseenko and G. V. Groshevoi, "Method of controlling the amplification and
       determining the frequency and amplitude response of seismic recording systems using
       an electromagnetic generator," Izv. Akad. Nauk SSSR, Ser. Geofiz., No. 5 (1953).

174.   S. A. Fedotov, "Some kinematic and dynamic properties of seismic waves refracted
       from curved interfaces," Tr. Sakhalin. Kompl. Nauchn.-Issled.Inst., No. 10, Izd. AN
       SSSR (1961).

175.   S. A. Fedotov, A. M. Bagdasarova, I. P. Kuzin, and R. Z. Tarakanov, "On the seis-
       micity and deep structure of the south Kuriles Island chain," Dokl. Akad. Nauk SSSR,
       Vol. 153, No. 3 (1963).

176.   S. A. Fedotov, N. N. Matveeva, R. Z. Tarakanov, and T. B. Yanovskaya, "On the
       velocity of longitudinal waves in the upper mantle in the region of the Japanese and
       Kuriles Islands," Izv. Akad. Nauk SSSR, Ser. Geofiz., No. 8 (1964).

177.   V. V. Fedinskii and Yu. V. Riznochenko, "Studies of the earth's crust," Vestn. Akad.
       Nauk SSSR, No. 6 (1962).

178.   E. E. Fotiadi and G. I. Karataev, "Crustal structure in Siberia and the Far East from
       the data of regional geophysical investigations," Geologiya i Geofizika, SO Akad. Nauk
       SSSR, No. 10 (1963).

179.   N. I. Khalevin and F. F. Yunusov, "On the use of elastic waves from industrial explo-
       sions for sounding of the earth's crust in the Urals," Izv. Akad. Nauk SSSR, Ser. Geo-
       fiz., No. 11 (1962), pp. 1567-1573.

180.   N. I. Khalevin and I. F. Tavrin, "On sub-horizontal layering of the upper part of the
       earth's crust in the Urals," Izv. Akad. Nauk SSSR, Fiz. Zemli, No. 3 (1965).

181.   N. I. Khitarov and A. B. Slutskii, "Effect of pressure on the melting temperature of
       albite and basalt," Geokhimiya, No. 12 (1965).

182.   D. Hughes and R. MacEwen, "Density of basic rocks at very high pressures," in:
       Dinamicheskie issledovaniya tverdykh tel pri vysokikh davleniyakh, Izd. Mir (1965).

183.   B. S. Chekin, "On the reflection of elastic spherical waves from an inhomogeneous
       halfspace," Izv. Akad. Nauk SSSR, Ser. Geofiz., No. 5 (1964).

184.   A. V. Chekunov and G. M. Pustovalova, "Use of near-critical reflections in deep
       seismic sounding on the southern slope of the Ukrainian Shield," Izv. Akad. Nauk
       SSSR, Ser. Geofiz., No. 2 (1964).

185.   V. F. Chervini, A. M. Epinat'eva, and I. P. Kosminskaya, "Features of reflected and
       head waves in the vicinity of the initial point," Izv. Akad. Nauk SSSR, Fiz. Zemli,
       No. 8 (1965).

186.   Yu. M. Sheinmann, "Possible relation of magma with the structure of the outer shells
       of the earth," Trans. of the Conference "Chemistry of the Earth's Crust," Vol. II, Izd.
       Nauka (1964).

187.   Yu. G. Yurov, "Crustal structure in the Caucasus and isostasy," Sov. Geol., No. 91
       (1963).

188.   G. A. Yaroshevskaya and A. N. Fursov, "Experiences in recording deep waves using
        stations with intermediate magnetic recording," in: Voprosi metodiki glubinnogo
        seismicheskogo zondirovaniya, Izd. Nauka (1965) [Problems in Deep Seismic Sounding,
        Consultants Bureau (1967)].

189.   S. Balakrishna and G. Subrahmanyam, "Variation of ultrasonic velocities in charnock-
        ites," Current Science, p. 31 (1962).

190.   F. Birch, "The velocity of compressional waves in rocks to 10 kilobars," J. Geophys.
        Res., Pt. 1, Vol. 65, No. 4 (1960); Pt. 2, Vol. 65, No. 7 (1961).

191.   H. Bradner and I. Dodds, "Comparative seismic noise on ocean bottom and on land,"
        J. Geophys. Res., Vol. 69, No. 20 (1964).

192.   E. C. Bullard and D. T. Grigg, "The nature of the Mohorovičic discontinuity," Geo-
        phys. J., Vol. 6, No. 1 (1961).

193.   E. C. Bullard, "The Mohole," Endeavour, Vol. 20, No. 80 (1961).

194.   E. C. Bullard, "Deep structure of the ocean floor," Proc. Royal Astron. Soc., p. 265
        (1962).

195.   V. Cerveny, "The amplitude curves of the reflected harmonic waves around the criti-
        cal point," Studia Geophys. et Geodaet., Vol. 5, No. 4 (1961).

196.   V. Cerveny, "On the length of the interference zone of reflected and head waves
        beyond the critical point, and on the amplitudes of head waves," Studia Geophy. et
        Geodaet., Vol. 6, No. 1 (1962).

197.   V. Cerveny, "On the position of the maximum of the amplitude curves of reflected
        waves," Studia Geophys. et Geodaet., Vol. 6, No. 3 (1962).

198.   V. Cerveny, "Simplified relations for amplitudes of spherical compressional har-
        monic waves," Studia Geophys. et Geodaet., Vol. 7 (1963).

199.   V. Cerveny, "On the amplitude of reflected waves for some models of earth's crust,"
        Veröffentl. Inst. Bodendynamik und Erdbebenforschung Jena, No. 77, Berlin (1964).

200.   V. Cerveny and J. Jansky, "Amplitudes of body waves propagating in the earth's
        crust," ESC Proceedings of the Budapest Meeting, 1964, UJJ (1965).

201.   S. P. Clark and A. E. Ringwood, "Density distribution and constitution of the mantle,"
        Rev. Geophys., Vol. 2, No. 1 (1964).

202.   H. Closs and J. Labrouste (Editors), "Recherches seismologiques dans les Alpes
        occidentales ou moyen de grandes explosions en 1956, 1958 et 1960. Memoire collec-
        tive du Groupe d'études d' explosion alpines," Centre National de la Recherche
        Scientifique, Année Geophysique Internationale, Participation Française, Ser. 12,
        Seismologie Fasc. 2 (1963).

203.   H. Closs, "Geophysikalische Arbeiten in der sudlichen Nordsee," Deutsch. Geolog.
        Ges., Vol. 144, No. 1 (1962).

204.   H. Closs, "Sous-commission des explosions alpines," ESC Proceedings of the Buda-
        pest Meeting, 1964, UGGI (1965).

205.   G. Dohr, "Zur reflexionsseismischen Erfassung sehr tiefer unstetigkeitsflächen,"
        Erdöl und Kohle, 10(1):4 (1957).

206.   G. Dohr, "Über Beobachtungen von Reflexionen aus dem tieferen Untergrunde im
        Rahmen routinemässiger reflexions-seismischer Messungen," Z. Geophys., Vol. 25,
        No. 6 (1959).

207.   I. Ewing, "The mantle rocks," The Sea, Vol. 3 (1963).

208.   R. Fisher and R. Raitt, "Topography and structure of the Peru-Chile trench," Deep-
        Sea Res., Vol. 9, Pergamon Press (1962).

209.   G. F. Frantty, "The nature of high frequency earth-noise spectra," Geophysics, Vol.
        27, No. 4 (1963).

210.   L. Galfi and L. Stegena, "Deep reflections and the structure of the earth's crust in the
        Hungarian plain," Geofizikai Kozlemenyek, Vol. 8, No. 4 (1960).

211.   M. Hayakana and S. Balakrishna, "An explanation for the high ultrasonic velocity in Indian rocks," Geophys. Prosp., Vol. 9, No. 1 (1961).

212.   International Upper Mantle Project, Canadian Progress Report (1963).

213.   International Upper Mantle Project, USA Progress Report (1963).

214.   W. N. Jackson and L. C. Pakiser, Seismic Study of Crustal Structure in the Southern Rocky Mountains, Geol. Survey Res., 525-D (1965).

215.   L. R. Janson, "Crustal structure between Lake Mead, Nevada, and Mono Lake, California," J. of Geophys. Res., Vol. 70, No. 12 (1965).

216.   C. Knothe, "Tiefenseismische Arbeiten in DDR," Veröffentl. Inst. angew. Geophysik, Bergakademie Freiberg, No. 140 (1964)

217.   I. P. Kosminskaya, "On layering of the earth's crust," J. Geophys. Res., Vol. 68, No. 4 (1963).

218.   I. P. Kosminskaya, "Classification of crustal structures by seismic data," in: Europ. Seismol. Kommission, Veröffentp. Inst. Bodendyn. und Erdbebenforsch., No. 77, Jena (1964).

219.   I. P. Kosminskaya, "Crustal studies in the USSR during IGY," Ann. Inst. Geophys. Year, "Seismology," Pergamon Press (1965).

220.   I. P. Kosminskaya and Yu. V. Riznochenko, "Seismic studies of the earth's crust in Eurasia," in: Research in Geophysics, Vol. 2, Solid Earth and Interface Phenomena, Washington (1964).

221.   H. Labroust, Etudes de la croute terrestre faites pendant L'AGI par les seismologues Francais, Ann. Int. Geoph. Year, "Seismology," Pergamon Press (1965).

222.   H. J. Liebschier, "Deutungsversuche für die Struktur der tieferen Erdkruste nach reflexionsseismischen und gravimetrischen Messungen im deutschen Alpenvorland," Z. Geophys., No. 2 (1964).

223.   R. McConnell and J. Taggart-Cowan, Crustal Seismic Refraction Profiles. A Compilation. The University of Toronto. Inst. of Earth Science, Scientific Report No. 8 (1964).

224.   E. Mituch, K. Posgay, and L. Sedi, "Szelesszigü reflexion alakalmazása a kéregkutataslan," Geofizikai Kozlemenyek, Vol. 13, No. 2 (1964).

225.   St. Muller, A. Stirn, and R. Veés, "Seismic scaling laws for explosions on a lake bottom," Geophys., No. 6 (1962).

226.   P. O'Brien, "A note on normal incidence reflections from the Mohorovičic discontinuity," Geophys. Trans. Roy. Astron. Soc., Vol. 9, Geophys. Suppl., No. 5 (1965).

227.   L. Pakiser and J. Steinhart, "Explosion seismology in the Western Hemisphere," in: Res. in Geophys., Vol. 2, Solid Earth and Interface Phenomena, Washington (1964).

228.   L. Pakiser, "Gravity, volcanism, and crustal structure in the southern Cascade Range," California. Geol. Soc., Ann. Bull, Vol. 75 (1964).

229.   L. Pakiser, "The basalt-eclogite transformation and crustal structure in the western United States," Geol. Survey Res. Spec. Paper, 525b (1965).

230.   C. Prodehl, Auswertung von Refraktionsbeobachtungen im bayrischen Alpenvorland, Deutsche Geophys. Ges., No. 4 (1964).

231.   R. W. Raitt, The Crustal Rocks, The Sea, Vol. 3, New York (1963).

232.   R. W. Raitt, "Geophysics of the South Pacific," in: Research in Geophysics, Vol. 2, Solid Earth and Interface Phenomena, Washington (1964).

233.   L. Stegena, "The structure of the earth's crust in Hungary," Acta Geologica. A. Magyar Tudamanyos Akademia Foldani Kozlonya, Vol. 8, F. 1-4 (1964).

234.   L. Stegena, "Research of crustal structure in Hungary," Ann. of Intern. Geophys. Year, Vol. 30, Seismology, Pergamon Press (1965).

235.   J. S. Steinhart and R. P. Meyer, "Explosion studies of continental structure," Carnegie Inst. of Washington (1961).

236.   D. I. Stuart, J. C. Roller, W. H. Jackson, and G. B. Mangau, "Seismic propagation path, regional travel time, and crustal structure in the western United States," Geophysics, Vol. 29, No. 2 (1964).

237.   United Kingdom contributions to the Upper Mantle Project, The Royal Sci. British National Committee for Geodesy and Geophysics, London (1963).

238.   P. S. Veitsman and I. P. Kosminskaya, "Summary of special seismic crustal studies during the IGY," Ann. of Intern. Geophys. Year, Vol. 30, Seismology, Pergamon Press (1965).

239.   D. Willis, "Comparison of seismic waves generated by different types of source," Bull. of Seism. Soc. of Am., Vol. 53, No. 5 (1963).

240.   P. Woollard, "Problems of the upper mantle and Hawaii as a site for the Mohole," Pacif. Sci., Vol. 19, No. 3 (1965).

241.   P. Wyllie, "The nature of the Mohorovičic discontinuity, a compromise," J. Geophys. Res., Vol. 68, No. 15 (1963).